Cover image note:
The cover image is a visual illustration of a region in which structure has formed through local interactions of process and boundary. It is not intended as a depiction of the universe or of any particular domain. The image suggests that stable structure arises through such interactions wherever conditions permit, and that what lies beyond the visible region may be governed by the same generative principles. The image is offered for visual context only.

The format and layout of the image were designed by the author. The visual content was generated with the assistance of an artificial image-generation system.

First Edition-4
© 2026 Christopher Effgen
All rights reserved.
No part of this book may be reproduced or distributed in any form without written permission from the author, except for brief quotations used in reviews or critical analysis.
The Generative Structure of Emergence
Published in the United States of America

Table of Contents

Cover image and copyright..i
Table of Contents...ii
Introduction..vi
Preface for Readers..vii
Scientific Preface..viii

Part I The Universe

CHAPTER 1 — Why "How" Must Precede "What"........1

CHAPTER 2 — Process, Boundary, and Collapse..........6

CHAPTER 3 — The Nine Modes: A Minimal Generative Basis for Interaction...12

CHAPTER 4 — Collapse Families: A Universal Grammar of Transitions..17

CHAPTER 5 — Field-Collapse and the Early Universe...25

CHAPTER 6 — Object-Collapse: The Emergence of Matter...32

CHAPTER 7 — Law-Collapse: How Physical Laws Form from Repetition..38

CHAPTER 8 — Hypothesis- and Expectation-Collapse in Physical Systems...46

CHAPTER 9 — Room- and Rail-Collapse in Physics.....54

Part II The Emergence of Life

CHAPTER 10 — Life as E/A in a Structured Field.........62

CHAPTER 11 — Biological Object-Collapse: Proto-Cells..69

CHAPTER 12 — Biological Law-Collapse.................76

CHAPTER 13 — Hypothesis- and Expectation-Collapse in Evolution..83

CHAPTER 14 — Room- and Rail-Collapse in Living Systems..91

Part III The Development of Cognition

CHAPTER 15 — Field-Collapse in Sensation.............99

CHAPTER 16 — Object-Collapse in Perception.........107

CHAPTER 17 — Cognitive Law-Collapse................115

CHAPTER 18 — Hypothesis and Imagination...........123

CHAPTER 19 — Expectation, Experimentation, and Learning...131

CHAPTER 20 — Constructed Contexts and Habit Formation..139

CHAPTER 21 — Cognitive Rail-Collapse and the Formation of Identity..147

Part IV Culture

CHAPTER 22 — Shared Field-Collapse: Collective Attention...155

CHAPTER 23 — Social Object-Collapse..................163

CHAPTER 24 — Cultural Law-Collapse..................171

CHAPTER 25 — Cultural Hypothesis- and Expectation-Collapse..179

CHAPTER 26 — Constructed Cultural Boundaries......187

CHAPTER 27 — Cultural Rails: Tradition, Inertia, Innovation, and Drift...195

Part V Artificial systems

CHAPTER 28 — Making Collapse Visible in Machines..203

CHAPTER 29 — QC Implemented in Machines: Discoveries From the GAI Project................................213

CHAPTER 30 — Artificial Rails and Drift................222

CHAPTER 31 — Could Artificial Collapse Become Cognitive?..231

Part VI Science

CHAPTER 32 — QC and the Structure of Scientific Inquiry..241

CHAPTER 33 — QC and the Unity of Science…,........250

CHAPTER 34 — Predictions and Research Programs...261

CHAPTER 35 — Modal Mechanics: A New Scientific Discipline..270

Epilogue — The Next Collapse................................279

Appendix A — On Human and Artificial Collaboration in This Work..285

Structural Glossary..288

Introduction

The Book Ahead

What follows is an exploration of how collapse operates across domains:

- **Part I** begins with the universe
- **Part II** with the emergence of life
- **Part III** with the development of cognition
- **Part IV** with culture
- **Part V** with artificial systems
- **Part VI** with science itself

Each section uses the same modal architecture to reveal new forms of structure.

The universe collapses into stars.
Life collapses into form.
Mind collapses into meaning.
Culture collapses into pattern.
Artificial systems collapse into observable structure.
Science collapses into understanding.

And QC provides the structural language to describe all of it.

Preface for Readers

How to Approach This Book

This book is not written to persuade, argue, or provoke.
It is written to describe structure.

Across science, philosophy, and everyday explanation, we are accustomed to asking *what* exists: what particles compose matter, what organisms do, what minds think, what cultures produce. This book asks a different question: how anything becomes structured enough to be a "what" in the first place.

The chapters that follow develop this question carefully and deliberately. Concepts are introduced only when they are needed, and each new domain is approached using the same structural language. Nothing in the book depends on accepting claims about consciousness, metaphysics, or artificial intelligence. The argument proceeds by pattern, not by assertion.

Readers may approach the book in different ways. It can be read linearly, as a developmental arc from the earliest physical structures to scientific understanding itself. It can also be read selectively, returning to earlier chapters as later ones clarify their role. Both approaches are valid.

No prior knowledge of Quanta of Cognition is required. All necessary ideas are introduced within the text. What *is* required is patience with a mode of explanation that values structure over conclusion and process over proclamation.

This book does not tell you what to believe.
It asks you to observe how structure forms.

Scientific Preface

Scope, Commitments, and Methodological Restraints

This book applies Quanta of Cognition (QC) as a structural framework across physics, biology, cognition, culture, artificial systems, and scientific inquiry itself. Its central claim is methodological rather than metaphysical: that emergence across domains follows a shared generative grammar organized around process, boundary, and collapse.

Several boundaries should be stated clearly at the outset.

First, QC does not unify the sciences by reduction. No domain is derived from another, and no substrate is privileged. The unity proposed here is structural: the same collapse architecture appears wherever stable structure emerges, regardless of scale or material.

Second, collapse is used in a strictly generative sense. It does not denote failure, destruction, or measurement-induced selection. Collapse refers to the event by which multiplicity resolves into a configuration stable enough to persist and shape subsequent dynamics.

Third, this work is not a theory of consciousness, nor does it advance claims about subjective experience beyond those required for structural analysis. Where cognition is discussed, it is treated as an emergent domain governed by the same collapse grammar as physical and biological systems, but instantiated in a distinct substrate.

Fourth, artificial systems are treated as instruments, not as minds. Their value in this work lies in their legibility: collapse dynamics can be observed, perturbed, and analyzed with a clarity unavailable in natural systems. No claims are made about artificial understanding or agency.

Finally, the framework is intended to be falsifiable. If collapse cannot be distinguished from process or outcome, if boundary variation does not alter resolution dynamics, or if the proposed collapse sequences fail to appear across domains, the framework fails.

The chapters that follow proceed with these constraints in place. The goal is not to replace existing scientific theories, but to articulate the generative structure they implicitly share.

This book is offered as a contribution to method.

Part I The Universe

CHAPTER 1 — Why "How" Must Precede "What"

The Structural Foundation of Scientific Understanding

Science has long been concerned with what exists:

- what particles compose matter
- what forces govern the universe
- what organisms do to survive
- what neurons encode
- what societies produce

Beneath every *what* lies a deeper question:

How does anything become structured enough to be a "what" in the first place?

This is not a metaphysical diversion. It is a structural question. Every system science studies—from quarks to cultures—is already the outcome of a generative process. By the time we can measure it, something has already organized: potentials have narrowed into a stable form.

Quanta of Cognition (QC) begins at that generative level.

1.1 Science Describes Outcomes; QC Describes Origins

Scientific theories track patterns:

- planetary orbits
- molecular reactions
- biological behaviors
- cognitive processes
- social dynamics

These patterns are results—the echoes of deeper stabilizations that made their persistence possible.

QC provides a structural account of:

- how forms come to exist at all
- how patterns crystallize and persist
- how constraints arise and guide flows
- how later domains inherit structure from earlier ones

Science studies the footprints.
QC studies the gait.

1.2 Three Universal Roles in Emergence

Every emergent event can be described as the interaction of three **roles**:

- **Process** — that which unfolds, varies, or moves; the locally active tendency toward change.
- **Boundary** — that which constrains, shapes, differentiates, or channels process.
- **Collapse** — the event by which a particular interaction between process and boundary resolves into a configuration stable enough to persist.

These are not substances, entities, or mechanisms. They are **relational roles**. What occupies them differs by domain.

If process occurs without boundary, activity remains diffuse.
If boundary exists without process, nothing happens.
If interaction does not collapse, nothing endures.

This triad—process, boundary, and collapse—is the minimal architecture by which a *how* yields a *what*.

1.3 Collapse Is the Engine of Structure

A collapse occurs whenever:

- multiple possibilities are funneled into a single trajectory,
- process engages constraint and stabilizes,
- a system steps into a more persistent configuration,
- a retentive structure is carved out of flux.

Across domains the particulars differ; the structure does not.

We see collapses in:

- physics (mode excitations, particle formation),
- chemistry (nucleation and crystal growth),
- biology (proto-cell membranes that seal a metabolism),
- cognition (a percept snapping into focus),
- culture (a norm settling into practice).

Each is an instance of the same generative pattern: process meets boundary; their interaction resolves; the result persists and becomes available for further development.

1.4 A Generative Architecture Beneath the Sciences

QC does not unify physics, biology, cognition, or culture by reducing one to another. It unifies them by revealing a shared **generative grammar**.

Across substrates—

- a quantum field,
- a metabolic network,
- a neural hierarchy,
- a social group,
- an artificial model—

we observe the same skeleton:

1. a field of undifferentiated possibility,
2. localized activity within that field,
3. encounters with constraint,

4. collapse into stabilized configurations,
5. and, later, the emergence of patterns, contexts, and inherited rails.

The book will use this grammar as its structural language. A full treatment of later-stage roles appears in subsequent chapters; here we establish the minimal architecture that drives emergence.

1.5 Why "How" Must Precede "What"

What physics studies—particles, forces, symmetries—are stabilized collapses of early-universe dynamics.
What biology studies—organisms, genes, metabolisms—are stabilized collapses of chemical and environmental interaction.
What cognitive science studies—percepts, categories, decisions—are stabilized collapses within neural fields.
What anthropology studies—norms, rituals, institutions—are stabilized collapses across interacting individuals.

Each discipline studies what has already collapsed.

QC studies **how collapse happens**—how process meets boundary, how interaction resolves, and how retentive structure becomes the substrate for further emergence.

That is why *how* must precede *what*: it is the only way to account for how a world of durable structure—physical, biological, cognitive, cultural, and artificial—comes to be at all.

QC offers the structural language for the chapters that follow.

CHAPTER 2 — Process, Boundary, and Collapse

The Universal Roles of Emergence

Every scientific discipline studies systems that change, interact, and stabilize. But before any system can be described in terms of what it becomes, something more fundamental must already be in place: a way for change to be shaped and resolved into structure at all.

Across physics, chemistry, biology, cognition, culture, and artificial systems, three roles recur wherever durable structure emerges:

- **Process** — that which unfolds or varies.
- **Boundary** — that which shapes or constrains process.
- **Collapse** — the event by which a particular interaction resolves into persistence.

This triad is not metaphorical. It is the minimal grammar required for emergence.

QC begins with the recognition that **roles are universal**, while the **forms that occupy those roles are domain-relative**.

2.1 Process: The Role of Change

In every domain, something happens.

In physics, fields fluctuate and interactions propagate.
In chemistry, reactions proceed and recombine.
In biology, metabolic cycles operate and behaviors unfold.
In cognition, attention shifts and neural activity propagates.
In culture, interaction, communication, and coordination occur.
In artificial systems, inference unfolds and candidate states are generated.

In each case, *process* names the role played by whatever is locally unfolding. Process is not a substance or a specific entity. It is a **position in an interaction**.

Process alone, however, cannot produce form.

2.2 Boundary: The Role of Constraint

Wherever process occurs, it does not occur freely.

Process is shaped by limits, resistances, contrasts, affordances, and contextual conditions. These play the role of boundary.

Physical boundaries include gradients, symmetries, and conservation constraints.
Biological boundaries include membranes, regulatory mechanisms, and environmental limits.

Cognitive boundaries include contrasts, inhibitions, and attentional limits.
Cultural boundaries include norms, roles, expectations, and institutions.
In artificial systems, boundaries appear as architecture, objectives, prompts, and available context.

Boundary is not a thing in itself. It is the role that shapes process. The same form may occupy boundary or process roles depending on context.

2.3 Collapse: The Role of Resolution

When process encounters boundary, many outcomes are possible. Most interactions dissolve without consequence.

A collapse occurs when a particular interaction **holds**.

Collapse is the moment when:

- a subset of possibilities binds,
- interaction stabilizes,
- and a configuration becomes persistent enough to be reused, combined, or inherited.

Collapse is not destruction. It is resolution. It is how a system selects one path among many and thereby produces a *what* from a *how*.

2.4 Roles and Carriers Across Domains

It is essential to distinguish **roles** from the **forms that occupy them**.

In cognition, the roles of process, boundary, and collapse are occupied by experiential forms **directly accessible as lived experience to individuals**. These experiential occupancies were the original site through which the collapse grammar was discovered and articulated.

In artificial systems, the same roles are occupied by explicitly engineered and observable mechanisms: inference dynamics fill the process role, architectural and contextual constraints fill the boundary role, and convergence events fill the collapse role.

This equivalence is strictly structural and functional. It does not imply that artificial systems have experience, experience-equivalents, understanding, or phenomenology of any kind.

Cognition provides **phenomenological access** to the generative grammar.
Artificial systems provide **instrumented access**.
Other domains provide only **residual access** through the structures that persist.

The equivalence asserted here is therefore one of **role-filling under collapse**, not of experiential kind or ontological status.

2.5 From Roles to Modes

From repeated interactions of process and boundary arise the modes—the nine structural configurations that describe how collapse unfolds across development.

These modes are not additions to the triad. They are specific configurations of it, discovered first in cognition and generalized structurally across domains.

They form the alphabet of generative structure.
Collapse is the syntax that combines them.

2.6 Collapse Families and Development

Across domains, collapses recur in recognizable transition types: stabilization of fields, formation of objects, extraction of patterns, opening of possibilities, licensing of prediction, construction of context, and inheritance of rails.

QC refers to these as collapse families. They are not mechanisms, but universal structural transitions through which systems deepen.

2.7 Why This Grammar Matters

Science studies the outcomes of collapse.
QC studies the conditions that make those outcomes possible.

By identifying process, boundary, and collapse as roles—and by distinguishing role-filling from experiential access—QC provides a disciplined way to reason across domains without reduction or metaphorical slippage.

This chapter establishes the universal roles.
The chapters that follow trace how their configurations unfold across the developmental arc of the universe, life, cognition, culture, and artificial systems.

CHAPTER 3 — The Nine Modes: A Minimal Generative Basis for Interaction

The Structural Alphabet of Emergence

Every system that evolves, stabilizes, or generates structure does so through specific relationships between process and boundary. QC identifies nine such relationships — the **nine modes** — which form the minimal generative basis for all emergent structure.

These modes are not metaphors.
They are structural types, the way topology has structural shapes or logic has fundamental operators.

The nine modes arise naturally from the combinatorics of process and boundary.

3.1 Why Nine Modes?

There are only so many meaningful ways process and boundary can relate:

- process without boundary
- boundary without structure
- process pressed into boundary
- process stabilized into form
- form repeated into pattern
- pattern freed into possibility
- possibility tested against boundary
- context constructed
- structure stabilized into rails

These nine relationships are the complete set of generative configurations necessary for emergence.

Nothing is redundant.
Nothing is missing.
The structure is minimal, sufficient, and universal.

3.2 A/A — Undivided Field

A/A is undifferentiated existence.
No object, no constraint, no direction.

Examples:

- early universe symmetry
- pre-patterned chemical fields
- infant sensory flux
- unstructured social space before coordination

A/A is pure potential.
Collapse has not yet begun.

3.3 E/A — Process Without Boundary

E/A is process in an undefined field:

- fluctuations
- chemical cycles
- exploratory neural activity
- spontaneous group motion

E/A is trying, reaching, exploring.
It is motion without contour.

3.4 A/E — Boundary Without Form

A/E is emerging constraint:

- gradients
- proto-membranes
- perceptual contrast
- social cues

A/E shapes process but is not yet structured into objects.

3.5 E/E — Structured Form

E/E is collapse into stable form:

- particles
- proto-cells
- percepts
- roles

E/E is the first appearance of "things" — internal or external.

3.6 M/E — Pattern from Repetition

M/E is abstraction:

- conservation laws
- metabolic patterns
- learned categories
- cultural norms

M/E compresses many forms into regularity.

3.7 M/A — Possibility from Pattern

When patterns are freed from their original boundaries, they become portable possibilities:

- potential energy states
- genetic variation
- imagination
- myth

M/A is the generative mode of exploration.

3.8 E/M — Testing Possibility Against Constraint

E/M is prediction:

- physical transitions
- biological selection
- cognitive expectation
- coordinated group action

E/M collapses hypothesis into feedback.

3.9 A/M — Constructed Boundaries

A/M is context creation:

- phase regimes
- regulatory networks
- cognitive scaffolds
- institutions

A/M builds the conditions under which future collapses occur.

3.10 M/M — Rails

M/M is stabilized identity:

- physical constants
- genetic inheritance
- habits
- traditions
- inference shortcuts in AI

Rails carry history forward.

3.11 Why These Nine Modes Form a Complete System

Together, the nine modes describe:

- every developmental step
- every emergent pattern
- every stable structure
- every collapse sequence

They form a generative basis
for structure at every scale.

This chapter introduced the alphabet of collapse.
The next chapter examines the collapse families — the universal transitions between modes.

CHAPTER 4 — Collapse Families: A Universal Grammar of Transitions

The Structural Pathways Through Which Systems Become Structured

If the nine modes are the alphabet of emergence,
then the collapse families are the grammar —
the universal transition types through which systems move
from one structural configuration to another.

Across physics, biology, cognition, culture, and artificial systems,
these transitions occur with remarkable consistency.
The substrate changes.
The scale changes.
The mechanisms differ.
But the structural pathways do not.

Collapse families are the universal ways systems transform.

QC identifies **seven collapse families**,
each corresponding to one of the major developmental transitions
from undifferentiated potential to stable rails.

4.1 Why Collapse Families Are Necessary

Modes describe static relationships between process and boundary.
Collapse families describe how systems *move* between modes.

This distinction is essential.

Modes give us a vocabulary of states.
Collapse families give us a vocabulary of transformations.

Together they form a complete structural language of emergence.

4.2 Field-Collapse

A/A → E/A + A/E

Field-collapse is the first step in emergence:
the moment undifferentiated existence breaks into process and constraint.

In physics, this appears as symmetry breaking.
In chemistry, as gradients forming or heterogeneity appearing.
In cognition, as attention emerging from sensory flux.
In culture, as shared focus emerging from scattered awareness.

Field-collapse creates contrast.
It is the first *decision* the universe makes.

4.3 Object-Collapse

E/A → E/E

Object-collapse stabilizes motion into form.

In physics: particle formation, bound states, nucleation.
In chemistry: crystals, micelles, molecular structures.
In biology: proto-cells, differentiated structures.
In cognition: percept formation.
In culture: roles, signals, recognizable social objects.

Whenever a process becomes a "thing,"
object-collapse has occurred.

4.4 Law-Collapse

E/E → M/E

When forms repeat under similar conditions,
patterns emerge.

Law-collapse distills repeated object-collapses
into regularities:

- conservation laws in physics
- metabolic cycles in biology
- concepts in cognition
- norms in culture
- embeddings in AI systems

Pattern is compressed history.

Law-collapse is how systems learn from recurrence.

4.5 Hypothesis-Collapse

M/E → M/A

Hypothesis-collapse frees patterns from the boundaries that produced them.

In physics: potential states, energy landscapes.
In biology: genetic variation, phenotypic exploration.
In cognition: imagination, mental simulation.
In culture: myth, symbolic recombination.
In AI: generalization, synthetic possibilities.

Hypothesis-collapse generates possibility.

It is how systems explore the space of what could be.

4.6 Expectation-Collapse

M/A → E/M

Possibility becomes prediction.
Hypothesis meets constraint.

Expectation-collapse tests imagined structure against boundaries:

- selection tests variation
- prediction tests expectations

- group coordination tests shared plans
- model inference tests candidate structures

Expectation-collapse is the moment systems learn from the world they inhabit.

4.7 Room-Collapse

E/M → A/M

Room-collapse constructs new boundaries from repeated successful predictions.

In physics: phase transitions create new constraints.
In biology: regulatory systems arise.
In cognition: schemas, task sets, conceptual scaffolds.
In culture: institutions, rituals, rule systems.
In AI: task contexts, prompt frames, inferential scaffolding.

Room-collapse builds the environments in which future collapses occur.

4.8 Rail-Collapse

A/M → M/M

Rails are the deepest stabilizations a system can achieve — persistent, default patterns that guide behavior.

In physics: constants, stable interaction types.
In biology: genetic inheritance, developmental programs.

In cognition: habits, identity structures.
In culture: traditions, languages, stable institutions.
In AI: optimized inference pathways, recurrent modes of operation.

Rail-collapse is the culmination of the collapse arc.

Rails are how systems inherit their own history.

4.9 Collapse Families Across Scales

Each collapse family appears repeatedly at different levels:

- A star forms through object-collapse, but so does a percept.
- A tradition stabilizes through rail-collapse, but so does a reflex.
- A hypothesis is tested through expectation-collapse, whether in evolution or learning.

The scale changes.
The mathematics changes.
The substrate changes.

The structural transition does not.

4.10 The Developmental Sequence of Collapses

Collapse families do not occur randomly.
They follow a natural developmental arc:

1. field-collapse
2. object-collapse
3. law-collapse
4. hypothesis-collapse
5. expectation-collapse
6. room-collapse
7. rail-collapse

This sequence appears in:

- cosmic evolution
- early life
- child development
- cultural evolution
- scientific inquiry
- artificial systems

The universality of the sequence is one of QC's deepest findings.

4.11 Collapse Families as a Scientific Tool

By studying collapse families we gain:

- predictive power (what structure comes next)
- diagnostic clarity (where systems break down)
- cross-domain insight (analogous transitions across fields)
- a unified scientific vocabulary

- a framework for modeling emergence

Collapse families let us talk about
how things come to be
rather than only what they are.

With this grammar in place,
we can now examine collapse families in context,
beginning with the universe.

Chapter 5 begins this journey.

CHAPTER 5 — Field-Collapse and the Early Universe

The Structural Beginning of Form

Every system that develops must begin with a moment of differentiation — a transition from undivided potential to structured possibility. For the universe, this transition corresponds to **field-collapse**, the moment when an undifferentiated field becomes patterned, directional, and capable of supporting further structure.

QC does not replace cosmology.
It provides the generative logic beneath it.
Field-collapse is the structural form beneath symmetry breaking, heterogeneity, and the earliest emergence of physical distinction.

This chapter introduces how QC interprets the first steps in the universe's development.

5.1 The Undivided Field (A/A)

At the earliest stage of the universe, before particles, before forces as we recognize them, before spatial differentiation, the cosmos existed in a state of profound uniformity.

Physicists describe this as:

- a symmetric field
- a homogeneous state
- a vacuum with no preferred direction

- an energy configuration without contours
- a condition lacking distinct objects or boundaries

In QC terms, this is **A/A** — undivided potential.

A/A is not emptiness.
It is *unshaped existence*.

There are no boundaries yet,
because nothing distinguishes one region from another.
There is no process yet,
because nothing has a direction in which to unfold.

A/A is the beginning before beginnings.

5.2 The Pressure Toward Differentiation

Uniform states are unstable under the right conditions — a principle well recognized in physics and complexity science. Even the slightest perturbation can amplify, leading to structure.

In cosmology, this appears as:

- quantum fluctuations
- early instabilities
- inflationary perturbations
- shifting energy densities
- symmetry-breaking potentials

QC abstracts the pattern:

Even in a perfectly uniform field, interactions carry tensions that eventually require resolution.

This is the precursor to collapse.
It is the subtle stirring before form.

Differentiation does not require intention.
It requires imbalance.

5.3 Field-Collapse: A/A → E/A + A/E

Field-collapse is the structural moment when undivided potential becomes structured possibility.

A field-collapse produces:

- **E/A**: the first directional processes
- **A/E**: the first boundaries, asymmetries, or constraints

This is not two events, but a single transformation with two outcomes.
Process and boundary co-emerge.

In physical terms, this corresponds to:

- the appearance of gradients
- the breaking of symmetries
- the emergence of local differences in field values
- the formation of directional flows
- the earliest irregularities that seed structure formation

Field-collapse is the universe's first "decision."
It is how the cosmos begins to draw distinctions.

5.4 Why Field-Collapse Precedes All Other Structure

Every form in the universe — particle, atom, cell, idea — ultimately depends on some earlier field-collapse.

You cannot have:

- object-collapse without differentiated process
- law-collapse without repeated forms
- hypothesis-collapse without patterns
- expectation-collapse without constraints
- room-collapse without constructed boundaries
- rail-collapse without stabilized regimes

Field-collapse is the root of emergence.
It creates the first contours of possibility.

Once the universe has process and boundary,
it can begin to generate structure.

5.5 Proto-Boundaries: The Universe Learns to Shape Itself

After field-collapse, the universe contains **proto-boundaries** — early constraints that influence how processes unfold.

These may include:

- variations in density
- curvature differences
- local energy configurations
- early fluctuations in field strength
- proto-gradients that channel flows

These proto-boundaries are not yet objects,
but they serve the same function:

they shape process.

The universe begins to "behave differently"
from one region to another.

This is the birth of form.

5.6 Field-Collapse as a Structural Bridge to Physics

QC does not claim what specific symmetry broke,
what field value dominated,
or what inflationary scenario occurred.

Those questions belong to physics.

What QC contributes is structural clarity:

Whatever the physical details, the early universe underwent a field-collapse — a shift from undivided to differentiated existence.

This is the universal first step in emergence:

- differentiation
- boundary appearance
- the birth of contrast
- the creation of directionality
- the onset of process

QC simply names the structure beneath these phenomena.

5.7 Why Field-Collapse Matters for Understanding the Universe

Field-collapse gives scientists a way to frame early-universe transitions in structural terms:

- early evolution is a modal sequence
- symmetry breaking is field-collapse
- gradients are proto-boundaries
- initial flow is E/A
- early confined regions are A/E
- structure begins through collapse, not imposition

Understanding the generative structure provides clarity about later transitions:

- how particles form
- how forces differentiate
- how stability emerges
- how physical laws arise

Field-collapse is the first rung on the ladder of the universe's development.

5.8 Transition to Object-Collapse

Field-collapse generates the conditions necessary
for the next major transformation:

E/A → E/E:
when processes stabilize into persistent forms.

This is object-collapse,
the structural foundation of matter.

The universe's first objects emerge from this transition —
a topic explored in the next chapter.

CHAPTER 6 — Object-Collapse: The Emergence of Matter

How Process Stabilizes into Form

After field-collapse, the universe contains:

- directed processes (E/A)
- and proto-boundaries (A/E)

There is now movement with contour,
tension with direction,
but still no stable "things."

To obtain objects that persist — particles, atoms, bound states —
the universe must undergo the next transformation:

$E/A \to E/E$
a collapse of doing into being.

This is **object-collapse**,
the structural moment when process stabilizes into form.

6.1 What Object-Collapse Is, Structurally

Object-collapse occurs when:

- a process (E/A) operates within a field of constraints (A/E),
- the interaction becomes self-reinforcing rather than self-dissolving,

- and a coherent configuration emerges that resists immediate disruption.

In other words:

A pattern of movement becomes a stable structure.

This is:

- a bound state in physics,
- a proto-cell in biology,
- a percept in cognition,
- a social role or signal in culture.

In every domain, object-collapse is how "something" appears.

6.2 Physical Echoes of Object-Collapse

In physics, object-collapse corresponds to phenomena such as:

- **Decoherence**
  where quantum superpositions resolve into classical outcomes.
- **Bound-State Formation**
  where quarks bind into hadrons, nucleons into nuclei, electrons into orbitals.
- **Phase Stabilization**
  where particles settle into stable phases or configurations as the universe cools.
- **Symmetry Breaking Outcomes**
  where previously equivalent configurations become distinct and stable.

The core idea in each case is the same:

repeated interactions under constraint produce a configuration that persists.

Object-collapse names that structural event.

6.3 Why Particles Are Collapse Outcomes, Not Primitives

Physics often treats:

- quarks,
- electrons,
- photons,
- protons,
- atoms

as fundamental entities.

QC reframes this structurally:

- these are **persistent solutions** of deeper process–boundary dynamics,
- not primitive givens,
- but the stable results of earlier collapses.

This does not change physics.
It places it within a generative story:

The universe did not start with particles.
It collapsed into particles.

Matter is a record of successful object-collapses.

6.4 The Criteria for Object-Collapse in Physical Systems

Across domains, object-collapse satisfies four universal criteria:

1. **Compatibility**
   The process must fit the constraints.
   In physics: energy, symmetry, and conservation requirements must be met.
2. **Stability**
   The resulting form must withstand small disturbances.
   In physics: bound states must have sufficient binding energy.
3. **Internal Coherence**
   Components must form a mutually reinforcing configuration.
   In physics: force balances, consistent field configurations.
4. **Recurrence**
   The structure must be reproducible under similar conditions.
   In physics: particles and atoms appear identically throughout the cosmos.

These criteria define object-collapse in structural rather than domain-specific terms.

6.5 Nested Object-Collapses: Building the Hierarchy of Matter

Object-collapse is recursive.

Each successful collapse creates new building blocks that can undergo further collapses:

- quarks → hadrons
- hadrons → nuclei
- nuclei + electrons → atoms
- atoms → molecules
- molecules → macromolecules

Each level:

- inherits constraints from lower levels,
- introduces new boundaries,
- supports new forms of process,
- and permits new collapses.

This stacking is how matter becomes chemically rich, how chemical richness sets the stage for life, and how life eventually sets the stage for mind.

The hierarchy of matter is a history of collapses.

6.6 Why Object-Collapse Matters for QC

Object-collapse is the first step where the universe becomes:

- *countable* (you can talk about "one" particle),
- *composable* (objects can combine),
- *measurable* (properties can be assigned and compared),
- *modelable* (laws can reference separate entities).

Without object-collapse, there would be:

- no atoms,
- no chemistry,
- no stars,
- no planets,
- no physical scaffolding for life.

Object-collapse is the transition
from flux to form.

6.7 Transition to Law-Collapse

Once stable forms exist in abundance,
they begin to interact in consistent ways.

The universe can then undergo:

E/E → M/E
the collapse of repeated structures into regularities.

This is **law-collapse**,
where patterns across many object-collapses
become physical laws.

Chapter 7 takes up that next step.

CHAPTER 7 — Law-Collapse: How Physical Laws Form from Repetition

Patterns Distilled From Stability

Once the universe has passed through object-collapse and produced stable forms — particles, atoms, bound configurations — it gains the capacity to form **patterns** across those forms. These patterns are not imposed from outside. They arise because the same kinds of collapses occur again and again under similar conditions.

When repeated forms and interactions converge into predictable regularities, the universe undergoes the next structural transformation:

E/E → M/E

This is **law-collapse** — the transition from individual stable forms to patterns that hold across them.

Physical laws are not decrees.
They are the structural memory of the universe's repeated collapses.

7.1 What Law-Collapse Is, Structurally

Law-collapse occurs when:

- many object-collapses (E/E)
- recur under similar constraints
- producing consistent outcomes
- which then stabilize into general regularities (M/E)

A law, in QC terms, is not a rule the universe "follows."
It is the **pattern that emerges from history**.

Across many interactions, the universe distills:

- symmetry
- invariance
- conservation
- preferred transitions
- stable relationships

These regularities form the backbone of physics.

7.2 How Repetition Creates Regularity

Repetition is the engine of law.
This is true in every domain.

In physics:

- electrons behave identically because their collapse conditions are identical
- the same energy transitions produce photons of the same frequency
- stable orbital configurations recur for all atoms
- transitions follow predictable branching ratios

Patterns do not appear at random.
They appear because the **same collapse conditions** produce the **same outcomes**.

QC names the structural moment when:

- many E/E events
- compress into an M/E regularity

as **law-collapse**.

7.3 Noether's Theorem and the Structure of Law-Collapse

In physics, Noether's theorem connects:

- symmetries
  to
- conservation laws

QC parallels that relationship structurally:

- repeated object-collapses under a symmetry
  → produce an invariant pattern
  → recognized as a physical law

Translation:

- conservation laws arise from collapse regularities
- symmetries are the modal conditions under which collapse is stable

Noether gives the mathematics.
QC describes the generative structure beneath it.

7.4 Why Laws Behave as if They Are Fundamental

To the scientific mind, physical laws can appear:

- timeless
- universal
- foundational
- unquestionable

QC reframes this without diminishing their status:

Laws appear fundamental not because they were present at the beginning,
but because they are the most stable outcomes of the universe's early collapse history.

The deepest rails are those that have survived the entire developmental arc.

Laws are rails (M/M),
but they are rails that solidified so early and so strongly that we experience them as if they were given.

7.5 Law-Collapse Across Scales

Law-collapse does not end once atoms stabilize.
It continues at every level of complexity:

Chemistry

- periodic patterns
- reaction pathways
- preferred bonding arrangements

Biology

- metabolic circuits
- regulatory patterns
- protein-folding preferences

Cognition

- categories
- invariances
- perceptual groupings

Culture

- repeated social behaviors
- ritual patterns
- conventions and customs

At every scale, repeated collapse generates regularity.

Physical laws are simply the earliest and most entrenched examples.

7.6 Why Law-Collapse Makes the Universe Predictable

Once law-collapse occurs, systems gain:

- reproducibility
- predictability
- stability
- coherence
- mathematical form

This is why physics can be written in equations:

- the underlying regularities have become structured enough
  to support abstract representation.

The universe after law-collapse is not chaotic experiment —

it is a landscape with stable contours.

This stability becomes the platform on which life emerges.

7.7 The Universe "Learns" Its Behavior Through Law-Collapse

The word "learn" must be used carefully.
The universe is not conscious.
But structurally, law-collapse *resembles* learning:

- repeated exposure → pattern detection
- pattern → constraint

- constraint → predictability

QC describes this as:

- many object-collapses → one pattern
- pattern → future collapse expectations
- expectations → new stability regimes

In this sense:

the universe behaves as if it accumulated wisdom about how to remain stable.

Not metaphorically — structurally.

7.8 Law-Collapse as Foundation for the Next Stage

Once laws exist:

- the universe can *explore* the possibilities permitted by those laws
- new combinations become available
- new pathways open

This is **hypothesis-collapse** (M/E → M/A):

- patterns freed from their original contexts
- degrees of freedom available for exploration
- space for novelty emerges

Just as cognition uses imagination,
the universe uses potential.

The next chapter describes this transition and how patterns give rise to possibility.

CHAPTER 8 — Hypothesis- and Expectation-Collapse in Physical Systems

From Pattern to Possibility, and from Possibility to Prediction

Once the universe has stabilized patterns through law-collapse, it possesses something entirely new: **a landscape of possibilities**. Regularity does not merely constrain behavior; it *opens* the space of what can happen.

Structures can combine.
Events can follow multiple paths.
Systems can occupy alternative configurations.
Outcomes can vary yet remain lawful.

At this stage, the universe begins to *explore* the range of forms permitted by its laws. QC names this generative shift:

M/E → M/A
pattern → possibility
law → hypothesis

And once possibilities emerge, the universe undergoes the next transformation:

M/A → E/M
possibility → prediction
hypothesis → expectation

46

Hypothesis-collapse and expectation-collapse are the structural basis for novelty, selection, dynamics, and the unfolding of complexity.

8.1 Hypothesis-Collapse: When Patterns Become Portable

Law-collapse produces patterns that hold across many cases.
But patterns are still tied to the contexts in which they formed.

Hypothesis-collapse frees these patterns.

A pattern (M/E) becomes:

- a potential configuration
- a degree of freedom
- an alternative expression
- an "open possibility" within the system

This is **M/A**, the abstract, aexperiential mode where structure becomes adaptable.

In physics, this appears as:

- potential energy surfaces
- state spaces
- alternative configurations of fields
- branching paths in quantum dynamics
- degrees of freedom in mechanical systems
- open sets of microstates under fixed macrostates

The universe begins to behave as if it is "considering" possibilities —
not with intention, but structurally.

Hypothesis-collapse is the universe discovering its own latitude.

8.2 Why the Universe Needs Hypothesis Space

Without hypothesis-collapse, the universe would be:

- static
- overly constrained
- devoid of complexity
- locked into a single path
- incapable of producing higher-order structure

Hypothesis space allows:

- atoms to form molecules
- molecules to fold into proteins
- proteins to form networks
- networks to generate metabolic cycles
- metabolic cycles to evolve into life

Every new structure depends on the system's ability to explore possibilities within the constraints of existing laws.

Hypothesis-collapse is the structural origin of novelty.

8.3 Expectation-Collapse: Testing Possibility Against Constraint

Once possibilities exist, they must be filtered.

Some configurations are stable.
Some pathways are favored.
Some transitions are probable.
Others are rare or forbidden.

Expectation-collapse is the moment when a system:

- applies pattern
- in a specific boundary
- to anticipate an outcome

In physics, expectation-collapse corresponds to:

- transition probabilities
- reaction likelihoods
- branching ratios
- decay pathways
- tunneling probabilities
- thermodynamic tendencies

In every case, the system "predicts" — structurally — where collapse is likely to land.

This is not cognition.
It is the modal logic beneath prediction itself.

8.4 Probability as Structural Expectation

Probability in physics reflects:

- the shape of the hypothesis space,
- the boundary conditions applied,
- and the structural constraints of the system.

Expectation-collapse formalizes this:

- M/A = the space of possibilities
- A/E = boundary conditions
- E/M = constrained expectation
- E/E = collapse into a specific outcome

What quantum physics calls "probability amplitudes," QC interprets as a structured interaction between:

- pattern,
- possibility,
- constraint,
- and collapse.

This does not conflict with physics.
It clarifies why probabilistic descriptions appear at all.

8.5 Hypothesis- and Expectation-Collapse Drive Complexity Forward

Through repeated cycles of:

- possibility (M/A)
- testing (E/M)
- collapse (E/E)

the universe develops increasing complexity.

Examples:

- molecular conformations that fold into the most stable structures
- reaction networks choosing among pathways
- energy landscapes guiding particle transitions
- structural configurations favoring low-energy states
- early chemical cycles stabilizing through repeated outcomes

The universe explores possibilities
and selects those that hold.

This is the structural precursor of evolution.

8.6 The Universe "Tests" Without Intention

Expectation-collapse can resemble decision-making:

- some outcomes succeed,
- others fail,
- successful configurations persist.

But this is not cognition.
It is **constraint satisfaction through collapse**.

Expectation-collapse is:

- lawful,
- structural,
- and substrate-neutral.

It is how systems resolve indeterminacy
into stable outcomes.

In life, this becomes biological selection.
In mind, it becomes learning.
In culture, it becomes coordination.
In AI, it becomes prediction.

In physics, it is the root of dynamical behavior.

8.7 Hypothesis Space and Cosmological Creativity

Hypothesis-collapse is responsible for the richness of structure we observe today:

- the diversity of atoms
- the range of chemical bonds
- the architectures of molecules
- the pathways of reaction systems
- the complex webs of chemistry that made life possible

These were not given.
They were *discovered* by the universe
through collapses in hypothesis space.

The more patterns existed,
the more possibilities emerged.

The more possibilities emerged,
the more collapses could be tested.
This recursive expansion is the deep engine of complexity.

8.8 Expectation-Collapse Sets the Stage for Room-Collapse

Expectation-collapse produces:

- stable expected outcomes
- stable preferred pathways
- stable interaction tendencies

When these expectations become so reliable
that the system reorganizes itself around them,
a new collapse emerges:

E/M → A/M
prediction → constructed constraint

This is **room-collapse**,
the formation of new boundary conditions —
phase transitions, stable regimes, structured environments.

Before the universe can build life,
it must build the physical "rooms" life requires.

Chapter 9 addresses this next transformation.

CHAPTER 9 — Room- and Rail-Collapse in Physics

How the Universe Builds Its Own Constraints and Stabilizes Them

By the time the universe has passed through hypothesis- and expectation-collapse, it has:

- stable forms (from object-collapse),
- stable patterns (from law-collapse),
- and a rich field of possibilities (from hypothesis-collapse),
- filtered through constraints (via expectation-collapse).

But possibility alone cannot build a universe that endures. For stability on cosmic scales, the universe must learn how to **organize itself**, shaping conditions under which collapses are reliable, repeatable, and robust.

This requires two new transformations:

Room-collapse (E/M → A/M)
and
Rail-collapse (A/M → M/M).

These collapses allow the universe to construct:

- new boundary conditions,
- new stable regimes,
- and deeply rooted regularities that act like physical "rails."

This chapter describes these two transformations at the level of physics.

9.1 Room-Collapse: When Expectations Build New Boundaries

Room-collapse occurs when the universe:

- repeatedly generates similar expectations (E/M),
- under conditions that strongly constrain outcomes,
- until the environment itself reorganizes into a stable configuration.

This is the physics behind:

- phase transitions,
- crystallization,
- domain formation,
- emergent field regimes,
- new stable interaction modes.

Room-collapse is the universe constructing **rooms** — contexts —
that support stable behavior.

These are not literal rooms,
but structured boundary conditions.

9.2 Examples of Room-Collapse in Physics

55

Phase Transitions

When temperature drops below a threshold, interactions reorganize:

- liquids freeze,
- gases condense,
- superconductivity emerges,
- magnetism appears.

These transitions create **new boundary conditions** — new regimes that constrain motion and interaction.

Crystallization

Atoms repeatedly encounter each other under low-energy conditions.
Their expectations (E/M) of preferred distances and angles shape the formation of lattices (A/M).

The solid is a constructed room:
an environment built from repeated collapse.

Emergent Field Regimes

When symmetries break and expectations stabilize, fields fall into specific configurations:

- the Higgs field's vacuum expectation value
- distinct force carriers
- quantized energy regimes

In QC terms, these are room-collapses.

9.3 Why the Universe Must Construct Its Own Boundaries

Without room-collapse, the universe would:

- remain chemically chaotic
- lack stable phases
- fail to support long-term structure
- collapse into triviality or heat death

Room-collapse creates **stability** on which higher-level phenomena depend:

- solid matter
- planetary formation
- chemical complexity
- biological viability

It is the physics equivalent of regulatory systems in biology and schemas in cognition.

Room-collapse is how the universe sets the stage for life.

9.4 Rail-Collapse: When Constructed Boundaries Become Deep Stability

Once constructed boundaries (A/M) persist across:

- many interactions,
- wide contexts,
- cosmic timescales,

they undergo **rail-collapse (A/M → M/M)**.

Rail-collapse creates:

- physical constants,
- stable interaction types,
- reliable symmetries,
- quantized properties,
- long-standing phase regimes.

Rails are the universe's deepest inheritances — structures that remain stable across billions of years.

They are not imposed laws.
They are the most entrenched outcomes
of the universe's collapse history.

9.5 Examples of Rail-Collapse in Physics

Fundamental Constants

Quantities like:

- the fine-structure constant,
- the gravitational constant,
- electron mass,
- charge quantization

are rails: deeply stabilized outcomes of earlier collapses.

Stable Interaction Categories

Electromagnetism, the weak force, and the strong force persist because their collapse pathways stabilized early.

Quantized Energy Levels

Quantum mechanics' discrete transitions
are rails formed by repeated, stable boundary conditions.

Cosmic Large-Scale Structure

The persistence of galaxies, clusters,
and cosmic web patterns
reflects deeply stabilized collapse regimes.

Rails are the universe's long-term memory.

9.6 Drift and the Limits of Rail Stability

Rails, while stable, are not absolute.
They can drift:

- under extreme conditions,
- during symmetry-breaking transitions,
- near black holes,
- in high-energy regimes,
- or in early-universe epochs.

Drift occurs when:

- new room-collapses override old constraints,
- rails become incompatible with new conditions,
- or boundaries fail under stress.

These events are rare today,
but QC predicts they were common at the universe's birth.

Rail-collapse is stability earned,
not stability decreed.

9.7 Why Room- and Rail-Collapse Matter for QC

These collapses create:

- stable matter
- stable chemistry
- stable environments
- reliable physical structure
- a universe capable of supporting higher-level collapse sequences

They are the structural joints between physics and the rest of existence.

If field-collapse sparks differentiation,
and object-collapse sparks structure,
and law-collapse sparks pattern,
and hypothesis-collapse sparks possibility,
and expectation-collapse sparks filtering,

then room- and rail-collapse spark **enduring stability**.

This stability is the foundation for:

- life,
- cognition,

- culture,
- artificial systems.

Without rails, there is no platform for evolution or thought.

9.8 Transition to Biology

With room- and rail-collapse complete,
the universe contains:

- stable phases,
- stable atoms,
- stable molecules,
- energy gradients,
- catalytic surfaces,
- planetary environments.

These are the **rooms** and **rails** on which life will emerge.

The next part of the book takes up this story:
how process and boundary give rise to living systems.

Part II The Emergence of Life

CHAPTER 10 — Life as E/A in a Structured Field

The First Biological Attempts

By the time physics completes its major collapses, the universe contains:

- stable matter (from object-collapse),
- stable patterns (from law-collapse),
- a landscape of possibilities (from hypothesis-collapse),
- and reliable constraints (from expectation-, room-, and rail-collapse).

These structures create **rooms** in which life can appear — not because life is inevitable, but because the universe has become rich enough in gradients, surfaces, cycles, and regularities to support sustained process.

Life begins not as organism, information, or replication, but as **process**.

10.1 The Physical Rails That Make Life Possible

Through billions of years of collapse, the universe prepares conditions suitable for life:

- stable atoms and molecules
- repeating chemical behaviors
- persistent energy gradients
- catalytic surfaces
- planetary cycles
- oceans, tides, and atmosphere
- temperature and pH ranges within viable bounds

These are **rails** — stabilized constraints that make complex chemistry possible.

Life does not arise despite the universe's structure.
Life arises **because of it**.

The physics of existence is the platform on which biology stands.

10.2 Life Begins as Process (E/A)

Before DNA, before membranes, before replication, life begins as **E/A** — process without autonomous boundaries:

- autocatalytic reaction cycles
- surface-bound molecules
- energetically driven flows
- oscillatory chemical systems
- redox loops and primitive metabolic patterns

These processes are not living,
but they exhibit the first hints of self-sustaining behavior.

They persist.
They draw in resources.
They alter local environments.
They repeat.

Life's earliest signature is not replication,
but the persistence of structured process.

10.3 Boundary Without Life: Environmental A/E

Early Earth provides **A/E** — boundaries the first biological processes borrow rather than build:

- mineral surfaces
- pores in rock
- hydrothermal vent structures
- clay lattices
- lipid films forming spontaneously
- cycles of wetting and drying
- gradients created by tidal and geothermal rhythms

These boundaries:

- concentrate molecules
- slow down diffusion
- enable organization
- channel reaction pathways
- stabilize short-lived intermediates

Boundary comes from the environment
long before organisms can generate their own.

Nature scaffolds life's first collapses.

10.4 Field-Collapse in the Prebiotic World

When environmental boundaries (A/E) combine with persistent processes (E/A),
life's first **field-collapses** occur at microscopic scales:

- reaction networks become partially confined
- gradients lock processes into recurring loops
- catalytic surfaces create proto-patterns
- compartments form transiently within pores or films

These are not yet cells,
but they are no longer mere chemical noise.

They are structured interaction zones —
proto-biological "rooms" built by geology and chemistry.

Biology begins in these niches.

10.5 Why Life Does Not Begin with Replication

A common misconception in origin-of-life thinking is that life begins when a molecule discovers how to copy itself.

But replication without self-maintenance is meaningless.
A copy is irrelevant if the system cannot persist long enough to benefit from it.

Before inheritance, before evolution, before information:

- there must be **coherence**
- there must be **sustained interaction**
- there must be **process–boundary coupling**

Life begins with **stability**,
not replication.

Replication is a later collapse.

10.6 The First Biological Object-Collapses (E/A → E/E)

When persistent chemical processes become strongly coupled to local boundaries,
a new collapse becomes possible:

E/A → E/E
process → coherent biological object

This is the structural moment of the **proto-cell**.

A proto-cell is not alive in the modern sense.
But it has:

- a semi-permeable boundary
- internal reactions that reinforce that boundary
- processes that maintain internal conditions

- the beginnings of autonomy
- the ability to persist through time

Boundary sustains process;
process sustains boundary.

A new kind of stability appears in the world.

10.7 Why Proto-Cells Are Necessary for Life

Life cannot exist without **persistent process-boundary units**.

Proto-cells provide:

- containment,
- internal gradients,
- catalytic cycles,
- self-maintenance,
- structural coherence.

Replication only becomes meaningful **after** proto-cells establish a context in which information can be preserved, variation can arise, and selection can operate.

Life's initial success is the success of **object-collapse**, not inheritance.

Protobiology is the story of process becoming form.

10.8 Nested Collapses and the Threshold of Evolution

Once proto-cells exist,
they become the environment for further collapses:

- **law-collapse** inside metabolism
- **hypothesis-collapse** inside molecular variation
- **expectation-collapse** through selection
- **room-collapse** as membranes improve regulation
- **rail-collapse** as early genetic systems stabilize

Each collapse adds structural depth
and increases the capacity for further collapse.

Life's development is a modal recursion.

10.9 Transition to Biological Law-Collapse

With proto-cells established through object-collapse,
the next transformation is inevitable:

$E/E \to M/E$
repeated biological behavior $\to$ stable metabolic patterns

Metabolic regularities become life's first laws.

Chapter 11 explores this next step.

CHAPTER 11 — Biological Object-Collapse: Proto-Cells

How Process and Boundary Become the First Living Forms

Life does not begin with genes, information, or replication.
It begins when chemical processes and environmental boundaries
collapse into **persistent units of self-maintenance**.

This transition — from process to coherent biological object —
is the biological expression of:

$E/A \to E/E$
process → form

The proto-cell is the first living object.
Not alive by modern standards,
but alive enough to shape the future of the planet.

11.1 Why Process + Boundary Is Not Yet Life

Before object-collapse, early Earth contains:

- vigorous chemical cycles (E/A)
- environmental boundaries (A/E)
- transient compartments
- osmotic gradients
- catalytic surfaces

These interactions are lively, structured, and sometimes persistent,
but they lack the key property of life:
internal coherence.

Life requires:

- a boundary that process sustains
- a process that boundary enables
- a mutually reinforcing loop

This coherence arises only when E/A collapses into E/E.

Before proto-cells, chemistry reacts.
After proto-cells, chemistry **persists**.

11.2 Environmental Boundaries as Scaffolds (A/E)

Early Earth is rich in boundary conditions:

- pores in volcanic and sedimentary rock
- lipid films and micelles
- mineral surfaces acting as catalytic basins
- iron-sulfur surfaces in hydrothermal vents
- clay lattices concentrating organic molecules
- tidal cycles generating dehydration–rehydration rhythms

These boundaries:

- concentrate reactants
- reduce entropy locally

70

- stabilize fleeting intermediates
- channel reaction flows
- create proto-compartments

Boundary arrives before life.
Life borrows boundary before it can build its own.

Environmental A/E is the scaffold for biological E/E.

11.3 Proto-Metabolism: Process Pressing Into Constraint (E/A → A/E)

The earliest life-like processes are not informational but **metabolic**:

- reaction loops powered by environmental gradients
- redox cycles sustained by mineral catalysts
- oscillatory chemical patterns
- primitive energy-harvesting chains
- surface-bound reactions gaining partial autonomy

As these processes bump against constraints:

- they adapt,
- stabilize,
- and become more coherent.

These structured E/A processes are the precursors to object-collapse.

Life is a boundary event waiting to happen.

11.4 The Structural Requirements for Biological Object-Collapse

Object-collapse in biology occurs when four conditions meet:

1. Compatibility

Internal reactions must fit the environment: temperature, pH, ion concentrations, substrates.

2. Stability

Boundary and process must reinforce each other: membrane-forming molecules, energy capture, waste management.

3. Internal Coherence

Reactions must begin forming a **network** rather than isolated events:
feedback loops, proto-regulatory interactions.

4. Recurrence

The same kind of proto-cell-like configuration must reappear reliably across environments.

When these conditions align,
process collapses into a coherent biological object.

11.5 The First Biological Object-Collapse: The Proto-Cell

A proto-cell is:

- a semi-permeable boundary (lipid vesicle or mineral compartment)
- housing internally coherent reactions
- that maintain or strengthen that boundary
- allowing the system to persist through changing conditions

Examples include:

- lipid vesicles stabilizing metabolic loops
- mineral-walled compartments guiding reaction chains
- surface-bound networks thick enough to generate an internal milieu
- dynamically stable protocellular droplets

When boundary and process become mutually sustaining, the system crosses the threshold:

process becomes form

This is the biological object-collapse.

11.6 Why Proto-Cells Are Not Optional in the Origins of Life

Every major theory of origin-of-life
— RNA-first, peptide-world, metabolism-first, membrane-first —
implicitly depends on proto-cell formation.

Without object-collapse:

- replication is meaningless,
- selection cannot operate,
- evolution cannot proceed,
- information cannot stabilize,
- metabolism cannot persist,
- and life cannot differentiate from environment.

Proto-cells are the first "selves" in the universe —
not conscious selves,
but bounded, persistent entities.

11.7 Nested Collapses: How Proto-Cells Set the Stage for Evolution

Once proto-cells exist,
each one becomes the environment for further collapses:

- **law-collapse** inside metabolism
  (stable reaction networks)
- **hypothesis-collapse** in variation
  (structural diversity)
- **expectation-collapse** through selection
  (fitness landscapes)
- **room-collapse** in regulatory pathways
  (homeostasis, feedback)

- **rail-collapse** in genetic and metabolic inheritance (stability across generations)

Biology descends from collapse into collapse, each one building on previous stability.

Life is collapse stacked upon collapse.

11.8 Proto-Cells Are the Turning Point

With proto-cells established:

- stability becomes heritable
- process becomes structured
- boundaries become self-generated
- reactions acquire histories
- failures carry consequences
- successes propagate

The universe now contains objects that can participate in evolution.

Life begins here.

The next chapter follows this story into the emergence of biological regularities — life's first laws.

CHAPTER 12 — Biological Law-Collapse

How Living Systems Stabilize Patterns Into Life's First Laws

Once proto-cells appear, the universe contains its first **persistent biological objects** — systems that maintain internal coherence long enough for patterns to form across them. These patterns are not imposed from outside. They arise because similar proto-cellular interactions repeat again and again in similar environments.

This repetition drives the next major biological transformation:

E/E → M/E
stable biological events → biological regularities

This is **biological law-collapse**, where metabolic patterns, reaction networks, and regulatory motifs emerge from the condensation of repeated processes.

Biology begins to develop its own internal laws.

12.1 From Chemical Noise to Biological Regularity

Before object-collapse, chemistry can be noisy and inconsistent.
After proto-cells form, something changes:

- reactions happen inside bounded spaces
- internal conditions remain more stable
- processes repeat in similar ways
- outcomes become predictable
- certain pathways are favored

Stability + repetition = pattern.

Biological law-collapse is life's first abstraction.

12.2 Why Biological Law-Collapse Is Necessary

Life cannot evolve without:

- stable metabolic cycles,
- reliable reaction sequences,
- predictable pathways,
- consistent responses to gradients,
- patterns of behavior inside cells.

Random chemical behavior cannot sustain inheritance. Stable patterns are required for:

- selection
- replication
- regulation
- adaptation
- growth

Biological law-collapse provides the **platform for evolution**.

12.3 Metabolic Law-Collapse: Recurring Cycles Become Patterns

As proto-cells repeatedly navigate similar environments, their internal reactions settle into **common cycles**:

- energy-harvesting sequences
- redox networks
- primitive glycolysis-like chains
- amino acid synthesis loops
- lipid-generation pathways

These cycles:

- reinforce internal boundary maintenance
- support energy storage
- stabilize proto-cell structure
- repeat across individuals and environments

Repetition → pattern → biological law.

Metabolism becomes **predictable**.

12.4 Reaction Networks as M/E Structures

A reaction network is more than a set of reactions. It is a **self-reinforcing configuration** of:

- inputs

- outputs
- catalysts
- feedback loops
- control points

Reaction networks are biological M/E:

- stable, repeatable, inherited patterns
- that organize biological doings
- into reliable pathways

Law-collapse is how metabolism becomes **organized**.

12.5 Genetic-Like Behavior Before Genes

Biological law-collapse predates DNA.
Before information can be inherited symbolically,
it must be inherited physically.

Repeated collapses generate:

- consistent membrane compositions
- stable metabolic tendencies
- preferred reaction pathways
- recurrent structural motifs

This non-genetic inheritance is life's first memory.

Modern genomes sit atop
older layers of law-collapse.

Genes encode patterns that chemistry had already discovered.

12.6 Differentiation of Biological Laws Across Contexts

Different environments produce different stable patterns:

- hot vs cold
- acidic vs alkaline
- shallow pools vs deep vents
- metal-rich surfaces vs clay lattices

Each environment:

- supports some collapse pathways
- suppresses others
- shapes metabolic tendencies
- favors certain reactions

Biological law-collapse is context-dependent.
It is not a universal chemical script.
It is the structural foundation of ecological diversity.

12.7 Biological Regularities Are Not Perfect — They Are Robust

Unlike physical laws, biological laws:

- have exceptions

- shift over evolutionary time
- differ across species
- adapt to environment

But they persist because they are:

- robust,
- repeatable,
- energetically favorable,
- supported by selection,
- integrated into larger networks.

Biological law-collapse produces **strong-enough** regularities
to sustain life without requiring perfection.

This is what allows life to be both stable and innovative.

12.8 Biological Law-Collapse Prepares the Way for Variation

Once stable metabolic patterns exist,
life can safely explore variation.

Hypothesis-collapse in biology emerges through:

- molecular variation
- structural differences
- metabolic branching
- errors in replication (once replication exists)
- alternative reaction paths
- adaptive mismatches and recoveries

Biology begins to **experiment** internally.

Pattern makes possibility safe.

Law-collapse enables life to take risks.

12.9 Transition to Evolution: Expectation-Collapse in Biology

Biological law-collapse creates:

- stable forms
- stable patterns
- predictable behavior

These lay the groundwork for the next transformation:

M/E → M/A → E/M
variation → possibility → selection

Biological expectation-collapse is the root of evolution.

Chapter 13 examines this transition —
how life begins to test its own possibilities
against environmental constraints.

CHAPTER 13 — Hypothesis- and Expectation-Collapse in Evolution

How Life Explores Possibility and Selects What Works

Once biological law-collapse stabilizes metabolic patterns,
life gains something new and profound:
the capacity to generate **variation** that matters.

Patterns create reproducibility.
Reproducibility makes deviation meaningful.
Meaningful deviation creates the space for evolution.

Evolution is the biological manifestation of:

M/E → M/A → E/M

- stable patterns → variations (hypotheses)
- variations → tested against constraints (expectations)
- successful outcomes → new biological form
- unsuccessful ones → collapse and disappearance

This chapter examines these two collapses — hypothesis and expectation —
as the engine of biological evolution.

13.1 Variation as Hypothesis-Collapse (M/E → M/A)

Variation is not random noise.
It is **structured departure** from established patterns.

Law-collapse creates stable metabolic rules.
Hypothesis-collapse frees those rules to be tested in new ways.

In biology, hypothesis-collapse includes:

- chemical variation in proto-cells
- early membrane differences
- metabolic branching
- enzymatic promiscuity
- structural diversity in molecules
- eventually, genetic mutation and recombination

Variation is the biological system
projecting patterns into the space of possibility.

It is imagination at the molecular scale.

13.2 Why Variation Is a Structural Necessity

Without variation:

- evolution cannot proceed
- novelty cannot emerge
- adaptation is impossible
- biological diversity would collapse
- ecological niches would remain empty

Variation is not an accident.
It is a structural inevitability in systems that have:

- pattern (M/E)
- persistence
- reproduction (eventually)
- environmental heterogeneity
- opportunity for divergence

Hypothesis-collapse is biology exploring what else is possible.

13.3 Expectation-Collapse: Selection as Modal Testing (M/A → E/M)

Once variation exists,
it must be filtered by reality.

Expectation-collapse in biology is **selection**,
the testing of biological possibilities
against environmental constraints.

Selection is not merely:

- competition
- survival pressure
- lineage sorting

It is the deeper collapse where:

- structural variation
- meets boundary conditions
- and only certain forms persist

Expectation-collapse is the bridge between possibility and outcome.

13.4 The Environment as Constraint: A/E in Evolution

Biological possibilities meet environmental boundaries when:

- temperature changes
- resources fluctuate
- predators appear
- toxins accumulate
- niches open or close
- physical environments shift
- competition arises
- ecological pressures reorganize

Each boundary condition tunes selection pressure differently.

Expectation-collapse is not a uniform filter.
It is a variable collapse shaped by A/E in each context.

Different environments → different evolutionary collapses.

This is why life radiates into diverse forms.

13.5 What Makes Selection a Collapse Event

A collapse event must:

- reduce many possibilities to one outcome
- have structural consequences
- support persistence
- shape future collapses

Selection does exactly this.

Variation creates branching pathways.
Selection collapses those branches into a stable lineage.

Successful forms become:

- ancestors
- founders
- templates
- rails for future evolution

Unsuccessful forms vanish.

Evolution is a collapse of biological hypothesis into biological expectation.

13.6 Repetition Makes Adaptation: Law-Collapse Returns

Surviving variations repeat across generations.
These repetitions become new patterns — a second-order biological M/E.

Adaptation is not a single collapse event.
It is a **nested collapse sequence**:

1. stable biology (M/E)
2. variation (M/A)
3. selection (E/M)
4. successful form (E/E)
5. repetition (back to M/E → new biological law)

Evolution is recursive:

- hypothesize
- test
- stabilize
- repeat

This recursive cycle can generate complexity,
not by design,
but by structural accumulation.

13.7 Why Evolution Is a Modal Structure, Not a Mechanism

Evolution is often described mechanistically:

- mutation + selection
- random change + filtering
- genetic variation across generations

QC reframes it structurally:

- M/E → stable biological form
- M/A → variation that expands possibility
- E/M → reality tests variation
- A/M → organisms build internal and external constraints

- M/M → inheritance stabilizes successful outcomes

Evolution is a **modal flow**,
not a merely genetic formula.

It is collapse sequencing in a living substrate.

13.8 Drift, Canalization, and Constraints as Modal Phenomena

Evolution exhibits more than selection:

Genetic Drift

Variation that persists without clear fitness differential:
M/A exploring neutral pathways.

Canalization

Developmental rails (M/M) that restrict what variation is viable.

Constraint-Breaking Innovations

New collapses that alter the entire collapse space
(e.g., multicellularity, nervous systems).

QC's modal vocabulary explains these phenomena without needing to stretch biological terminology.

13.9 Hypothesis- and Expectation-Collapse Prepare for Multicellularity

Variation and selection eventually produce:

- adhesion mechanisms
- coordinated signaling
- division of labor
- persistent cooperation
- stable developmental rails

These changes create conditions for the next great biological transformation:

room-collapse in living systems
(E/M → A/M in biology)

where organisms begin to build their own regulatory and developmental frameworks.

Chapter 14 continues with this transition.

CHAPTER 14 — Room- and Rail-Collapse in Living Systems

How Life Constructs Its Own Boundaries and Stabilizes Them

After evolution establishes variation and selection,
living systems gain the capacity to **shape the conditions under which future collapses occur**.
This is a profound shift: life begins not only to respond to constraints,
but to build them.

In QC terms, life enters the domain of:

Room-collapse (E/M → A/M)
and
Rail-collapse (A/M → M/M)

These transformations create regulatory networks,
developmental programs,
homeostatic systems,
and hereditary channels —
the structural backbone of biological organization.

14.1 Room-Collapse in Biology: Constructing Biological Boundaries

Room-collapse occurs when organisms or proto-organisms
build their own constraints — internal or external —
to stabilize processes that matter for survival.

Examples include:

- membranes becoming selective rather than passive
- ion pumps creating controlled gradients
- metabolic feedback creating regulatory domains
- early cytoskeletal elements organizing internal space
- molecular chaperones shaping protein folding
- compartmentalization within cells

In room-collapse:

- **E/M** (expectations about internal or external conditions)
  become
- **A/M** (constructed boundaries that enforce those expectations)

Life becomes a builder of rooms.

14.2 The Rise of Regulation

Life's earliest regulatory systems include:

- proto-homeostatic loops
- primitive sensory gating
- feedback inhibition
- early phosphorylation networks
- allosteric modulation

Each regulatory innovation:

- creates a new boundary,

- stabilizes previously volatile reactions,
- and increases internal coherence.

Room-collapse is how biology constructs
the conditions under which its processes remain viable.

14.3 Regulatory Rooms Are Developmental Milestones

As life becomes more complex,
room-collapse builds increasingly sophisticated structures:

In Single Cells

- cytoplasmic organization
- stress-response modules
- nutrient-sensing paths
- internal compartmentalization

In Early Multicellularity

- adhesion molecules
- extracellular matrices
- spatial patterning cues
- primitive developmental boundaries

In Complex Organisms

- tissues and organs
- hormonal signaling
- layered homeostatic controls
- functional specialization

Each new room influences which collapses can follow.

14.4 Rail-Collapse in Biology: Inheritance and Identity

Once constructed boundaries persist across conditions and generations,
they undergo **rail-collapse (A/M → M/M)** and become:

- conserved developmental programs
- gene regulatory networks
- metabolic rails
- canonical cell types
- reliable organismal structures
- reproductive strategies

Rails in biology are deeply stabilized:

- homeostatic set points
- conserved metabolic pathways
- consistent body plans (phylum-level rails)
- stable nervous system architectures
- inherited stress-response patterns

Rails are how life remembers.

14.5 The Dual Nature of Biological Rails

Rails both **enable** and **limit** evolution.

Rails enable:

- reliable reproduction
- stable development
- consistent physiology
- predictable behavior

Rails limit:

- phenotype space
- developmental plasticity
- possible innovations
- evolutionary direction

Rails are evolutionary commitments.

Once in place, they shape the future
by constraining which collapses remain possible.

14.6 Multicellularity as a Room-Collapse Masterpiece

Multicellularity represents a dramatic expansion of room-collapse:

- cells construct extracellular rooms
- developmental boundaries form
- gradients organize tissues
- signal pathways partition roles
- positional information emerges
- differentiation becomes stable

This new architecture enables:

- division of labor
- bodily stability
- immune systems
- nervous systems
- behavior

Room-collapse builds the body so that evolution can build the mind.

14.7 Nervous Systems as Rails for Cognition

Nervous systems exemplify the power of rail-collapse:

- sensory maps
- motor circuits
- reflex arcs
- synaptic stabilization
- developmental wiring plans

These are biological M/M patterns
that create **stable internal dynamics**
supporting cognition.

Before the mind can form percepts, categories, or concepts, biology must first build rails that support fast, reliable collapse.

Rails prepare biological systems
for the internalization of modal structure —
the topic of the next part of the book.

14.8 Why Room- and Rail-Collapse Are Prerequisites for Mind

Cognition requires:

- stable internal boundaries (A/M)
- reliable processing rails (M/M)
- persistent internal gradients
- repeatable input-output patterns
- a body capable of controlling its environment
- regulatory stability across time

Without biological rooms and rails:

- there is no platform for perception
- no basis for learning
- no architecture for memory
- no capacity for reflection
- no substrate for internal collapse

Mind is possible *because* biology mastered room- and rail-collapse.

14.9 Transition to Cognition

At this stage:

- life maintains boundaries,
- regulates itself,
- inherits structure,
- organizes internal space,
- and stores stable developmental information.

The system is now capable of a new transformation:

collapse begins to occur **inside** the organism.

The next chapter explores:

Field-collapse in sensation
and the birth of cognition.

Part III The Development of Cognition

CHAPTER 15 — Field-Collapse in Sensation

The Birth of Attention and the Origins of Mind

Cognition does not begin with thought, memory, or meaning.
It begins with **sensation** — the direct encounter between an organism and the world.
But sensation at its earliest stage is not structured.
It is undifferentiated experience, a continuous field with no objects, no distinctions, no boundaries.

QC describes this as **A/A** inside the nervous system:
a state of raw potential in which everything is present but nothing is separated.

For cognition to emerge, this experiential field must collapse.
It must differentiate into:

- **E/A:** directed activation (focus, proto-attention)
- **A/E:** internal boundaries (contrast, inhibition, suppression)

This is **cognitive field-collapse**, the first structural event in mind.

15.1 The Infant Mind as Undifferentiated Field (A/A)

Across species with nervous systems, early sensory activity appears as:

- diffuse neural activation,
- global excitability,
- uncoordinated firing patterns,
- unsegmented experiential flux.

Neuroscientists describe the earliest stages of perception as:

- unrefined thalamocortical dynamics,
- immature sensory maps,
- broadly tuned responses,
- undifferentiated cortical excitation.

Phenomenologically, this is the newborn's world:
a blur of intensity without delineation.

QC names this state **A/A**:
experience without structure.

There are no percepts yet.
No stable differences.
No "this" or "that."
Only the field.

15.2 Field-Collapse in Cognition: The First Differentiation

Field-collapse is the moment when undifferentiated sensation becomes structured.
In this collapse:

- some sensory activity becomes **foregrounded** (E/A)
- some becomes **backgrounded or inhibited** (A/E)

This is proto-attention.

Not selective attention as adults experience it,
but the very first emergence of focus and contrast.

A cognitive field can only become meaningful
once differentiation begins.

Field-collapse is the mind's first gesture toward structure.

15.3 The Emergence of Internal Boundaries (A/E)

In a differentiating sensory field, certain patterns of activation inhibit others.
This inhibition, whether in simple creatures or mammals, is the birth of **internal boundaries**.

Examples include:

- lateral inhibition sharpening visual edges
- gain control normalizing intensity
- early feature detectors tuning to simple contrasts
- primitive auditory segregation
- suppression of irrelevant or weak signals

These boundaries are not conceptual.
They are structural constraints created by neural architecture.

Internal A/E makes it possible for E/A to be meaningful.

Without boundary, focus cannot hold.

15.4 Directed Sensation (E/A): The Birth of Focus

Once internal boundaries form,
certain signals rise to prominence.

This is focus — the earliest form of directed experience.

Examples include:

- orienting reflexes
- preferential attention to caregiver faces
- focusing on movement
- gravitating toward high-contrast patterns
- responding to warmth, pressure, or voice

E/A is the first active stance the mind takes toward the world.
It is not conceptual or reflective,
but it creates the directional structure
from which all later cognition grows.

15.5 Why Field-Collapse Is Necessary for Perception

No system can perceive a "thing"
until it has differentiated the sensory field enough to notice:

- boundaries
- contrasts
- salience differences
- proto-objects

Field-collapse is not yet perception.
It is the precursor to perception.

Just as the universe collapses from undifferentiated fields
into directional process,
the mind collapses its sensory field
into the first glimmers of form.

Cognition recapitulates emergence.

15.6 Neuroscientific Echoes of Field-Collapse

Research in early sensory processing parallels QC's structure:

- **Thalamic gating:**
  early filtering and modulation
- **Oscillatory entrainment:**
  organizing raw input into temporal segments

- **Retinotopic and tonotopic maps:**
  early spatial ordering of sensory signals
- **Feature-based inhibition:**
  suppression creating perceptual edges
- **Competition for neural resources:**
  proto-boundaries creating proto-focus

These mechanisms generate the internal A/E structures that make E/A possible.

Neuroscience provides the mechanisms;
QC provides the structure.

15.7 Field-Collapse Sets Up Object-Collapse in Cognition

The goal of field-collapse is not differentiation for its own sake.
It is to prepare the mind for the next collapse:

E/A → E/E
the formation of percepts.

A percept is a stable internal object.
It is the cognitive equivalent of a particle or a proto-cell.

But percepts cannot form in a sensory field
that has not yet been shaped.

Field-collapse provides the contour and direction
that make perceptual collapse possible.

15.8 Why the Mind's First Collapse Mirrors the Universe's First Collapse

The parallel between cosmological and cognitive field-collapse
is not metaphorical.
It is structural:

- undifferentiated sensory field ↔ early cosmic symmetry
- emergent focus ↔ early fluctuations
- internal inhibition ↔ proto-boundaries
- organized sensory gradients ↔ early asymmetries
- proto-attention ↔ directional process

In both cases:

- collapse breaks uniformity,
- creates contours,
- sets the stage for stable objects,
- and defines the next mode.

Mind is not separate from nature.
It is a late expression of the same structural logic.

15.9 Transition to Object-Collapse in Perception

Field-collapse generates:

- contrast
- focus

- early internal constraints
- the first sense of "something, not everything"

The next transformation builds on this:

E/A → E/E
percepts forming from directed sensation.

This is object-collapse in cognition —
the moment the mind creates its first internal "things."

Chapter 16 takes up this next collapse.

CHAPTER 16 — Object-Collapse in Perception

How Doings Become Things Inside the Mind

After field-collapse organizes sensation into focus and contrast,
the mind is ready for its next transformation:
the emergence of perceptual objects.

Before this collapse, the organism has activity without structure,
sensation without recognition,
experience without stability.

To perceive, the mind must take a directed process (E/A)
and collapse it into a stable unit (E/E):

E/A → E/E
focus → percept

This is **object-collapse in cognition** —
the moment when the mind gives shape to the world.

16.1 Why Perception Requires Collapse

Perception is not passive reception.
It is an **active stabilization** of experience.

Raw sensory flux provides:

- stimulation
- gradients
- contrast
- focus

But percepts are:

- bounded
- persistent
- coherent
- meaningful
- recognizable

Perception requires a collapse event:
a resolution of competing sensory signals
into a single, stable internal form.

16.2 What Cognitive Object-Collapse Is

In cognitive object-collapse:

- focus stabilizes,
- internal boundaries sharpen,
- neural activity converges into an attractor state,
- and the mind "locks onto" a pattern that holds.

This collapse produces:

- the shape of a face
- the outline of a hand
- the sound of a voice
- a coherent movement

- a recognizable spatial region

These are not yet concepts.
They are percepts — cognitive objects.

Object-collapse is **the first appearance of "things" inside the mind.**

16.3 Neural Echoes of Object-Collapse

Neuroscience describes object formation through:

- **attractor dynamics:** stable neural activation patterns
- **recurrent processing:** top-down and bottom-up interactions converging
- **feature binding:** combining color, shape, motion, orientation
- **predictive coding stabilization:** minimizing error to produce a coherent percept
- **lateral inhibition:** sharpening edges and contours

These mechanisms are the physiological substrate of the structural collapse QC describes.

16.4 Percepts as Cognitive E/E

Percepts exhibit all the structural properties of E/E forms:

Stability

They persist long enough to serve as units of experience.

Boundaries

They differentiate themselves from background.

Coherence

Their internal structure fits together consistently.

Recurrence

They appear reliably under similar conditions.

Percepts are not images stored in the brain.
They are collapse outcomes —
momentary structured resolutions of sensory input.

16.5 Object-Collapse in Perception Is Recursive

Perception is not a single collapse but a cascade:

- a basic outline collapses
- then internal detail collapses
- then relational structure collapses
- then motion collapses into trajectory
- then identity collapses ("the same thing as before")

Perception is hierarchical, each collapse enabling the next.

The mind builds percepts layer by layer.

16.6 Why Object-Collapse Is the Foundation of Understanding

Perception does not merely present the world.
It *structures* it.

Without percepts:

- there can be no categories
- no concepts
- no learning
- no planning
- no communication
- no memory
- no meaning-making

Object-collapse is the first cognitive event
that creates something recognizable and stable enough
to become the basis for everything that follows.

It is the cognitive analogue of particle formation in physics
and proto-cell emergence in biology.

16.7 Percept Formation Is Not Recognition (Yet)

Percepts are:

- pre-conceptual,
- pre-linguistic,
- pre-categorical.

They do not yet involve:

- generalization
- abstraction
- meaning
- labeling
- classification

A percept is simply a structured "this."

Recognition — "this is a dog," "this is a cup," "this means danger" —
emerges at the next stage, when many percepts collapse into patterns (law-collapse).

Object-collapse prepares the raw materials of understanding.

16.8 How Perception Balances E/A and A/E

Object-collapse requires a delicate interplay between:

E/A (process)

Neural activation flowing through sensory pathways.

A/E (boundary)

Inhibition, contrast, constraint, structure.

Too much E/A → noise, instability, hallucination.
Too much A/E → rigidity, insensitivity, failure to pick up signal.

Perception is the equilibrium between exploration and limitation.

Collapse occurs when this interplay finds a coherent solution.

16.9 Object-Collapse Supports Memory and Learning

Percepts can become:

- stored as traces
- linked to outcomes
- compared across time
- used to form categories
- incorporated into expectations

Perception is not only immediate.
It is the substrate of memory:

- "I have seen this before."
- "This matches a known pattern."
- "This is different from what I expected."

Memory begins with perceptual collapse.

16.10 Transition to Conceptual Patterns (Law-Collapse)

Once percepts exist in abundance:

- the mind can detect regularities across them
- stable perceptual patterns become conceptual patterns
- categories emerge
- invariances take shape
- understanding becomes possible

The cognitive transition:

E/E → M/E
percept → pattern
form → concept

is the beginning of meaning.

Chapter 17 explores this next stage:
the emergence of conceptual structure through cognitive law-collapse.

CHAPTER 17 — Cognitive Law-Collapse

How Repeated Percepts Become Patterns, Categories, and Concepts

Once object-collapse generates stable percepts,
the mind contains its first internal "things."
But percepts alone do not provide understanding.
Understanding requires **patterns** across percepts —
structures that persist through variation, context, and time.

When repeated perceptual collapses converge into stable regularities,
the mind undergoes:

E/E → M/E
percept → pattern
form → concept

This is **cognitive law-collapse**,
the structural foundation of learning, meaning, and thought.

17.1 Why Percepts Are Not Enough

Percepts are:

- immediate
- local
- specific
- singular
- unintegrated

They tell the mind *what is happening right now*, but they do not tell the mind:

- what usually happens
- what goes together
- what belongs in a category
- how to anticipate
- what to ignore
- what matters

Cognitive structure begins only when percepts are distilled into patterns.

This distillation is **law-collapse** in the cognitive domain.

17.2 What Cognitive Law-Collapse Is

Cognitive law-collapse occurs when:

- multiple percepts
- sharing common structure
- under similar conditions
- produce a stable conceptual regularity

The mind begins to extract:

- invariances
- similarities
- recurring shapes
- co-occurrence tendencies
- relational patterns
- contextual consistencies

These produce:

- categories
- prototypes
- features
- conceptual boundaries
- foundational inferential structure

This is the mind's first abstraction.

17.3 Concept Formation as M/E

Concepts arise when:

- enough percepts share enough structure
- to justify treating them as "the same kind of thing"

The internal pattern (M/E) becomes:

- reusable
- recognizable
- generalizable
- stable
- referenceable

Concepts are the building blocks of symbolic thought.

But even without language,
cognitive systems rely on conceptual stability
to make sense of their environments.

Animal cognition, infant cognition, and human cognition
all operate with M/E structures.

17.4 Neural Signatures of Pattern Extraction

Neuroscience sees law-collapse in:

- Hebbian learning ("neurons that fire together wire together")
- strengthened synaptic pathways
- emerging feature detectors
- invariant object recognition
- temporal pattern learning
- predictive-coding hierarchies

These mechanisms convert:

- repeated percepts → stable representations

In QC terms, they implement cognitive M/E.

17.5 Law-Collapse Creates Cognitive Efficiency

Once patterns exist, the mind no longer needs to:

- reconstruct everything from scratch
- re-detect every perceptual feature
- analyze every signal with equal weight
- process every detail as if new

Patterns provide:

- shortcuts
- compression
- structure
- reduced cognitive load
- faster recognition
- better generalization

Understanding becomes *accumulative*.

17.6 The Difference Between Categories and Percepts

Percepts reflect:

- raw sensory structure
- moment-to-moment collapse

Categories reflect:

- stable regularities
- accumulated predictions
- boundary abstractions
- culturally reinforced refinements (in humans)

Percept:
"This particular shape is in front of me."

Category:
"This is a *kind* of thing I have encountered before."

Concept:
"I can operate on this category abstractly, even without seeing it."

All three depend on cognitive collapse,
but only law-collapse creates conceptual stability.

17.7 Why Law-Collapse Is the Foundation of Meaning

Meaning requires:

- stability (E/E)
- repeatability (M/E)
- connection across contexts
- the ability to relate current perception to past experience

Law-collapse provides all four.

Without conceptual patterns:

- there is nothing for expectations to act on
- predictions cannot form
- memory cannot organize
- learning cannot accumulate
- communication becomes impossible

Meaning is not a primitive.
It is a collapse outcome.

17.8 Pattern Extraction Feeds Hypothesis-Collapse

Once stable patterns exist,
they become the raw material for possibility:

M/E → M/A

The mind can:

- imagine variations,
- mix patterns,
- anticipate unseen combinations,
- form analogies,
- simulate future scenarios.

Hypothesis-collapse depends on conceptual stability.

Without M/E, imagination would drift without structure.

Patterns make exploration safe.

17.9 Conceptual Drift and Reclassification

Law-collapse is robust but not static.

Patterns can drift when:

- new percepts accumulate
- environments change
- cultures refine categories
- expertise develops
- expectations reshape boundaries

Cognitive drift is the shifting of conceptual M/E structures as new collapses accumulate.

Categories evolve.

Even meaning evolves.

This is a normal and healthy part of cognitive development.

17.10 Transition to Cognitive Possibility (Hypothesis-Collapse)

With conceptual patterns established:

- the mind can explore
- compare
- remix
- project
- question
- create

The next collapse in cognition is:

M/E → M/A
pattern → possibility
concept → imagination

This marks the beginning of hypothesis-collapse in the mind —
creativity, generalization, and internal generation of alternatives.

Chapter 18 continues that story.

CHAPTER 18 — Hypothesis and Imagination

How the Mind Frees Patterns From Context

Once the mind has formed stable conceptual patterns through cognitive law-collapse, it gains a new and transformative capacity:
the ability to **free** those patterns from the situations in which they were learned.

This is the beginning of **imagination**,
the internal exploration of possibility.

In modal terms, this is:

M/E → M/A
pattern → possibility
concept → hypothetical variation

Hypothesis-collapse opens a new dimension in cognition: the mind is no longer limited to what it has encountered. It begins to generate what *could* be.

18.1 What Hypothesis-Collapse Is in Cognition

A pattern (M/E) becomes a **portable hypothesis (M/A)** when the mind can:

- modify it
- project it

- combine it with others
- apply it in new contexts
- imagine it transformed
- simulate its consequences
- test it against internal or external constraints

Hypothesis-collapse is not fantasy or idle speculation.
It is the structural mechanism by which cognition becomes creative.

The mind begins to explore the space of forms permitted by its conceptual rails.

18.2 Why Hypothesis-Collapse Is Not Optional

Without hypothesis-collapse:

- there would be no imagination
- no creativity
- no analogy
- no scientific reasoning
- no planning
- no counterfactual thinking
- no flexibility
- no problem-solving beyond direct experience

Hypothesis-collapse is the primary engine of adaptive intelligence.

It allows the mind to prepare for the future
without needing to live through every possibility.

18.3 Neural Correlates of Hypothesis-Collapse

Brain networks that support imagination include:

- the **default mode network**, associated with internal simulation
- the **prefrontal cortex**, supporting abstraction and rule recombination
- the **hippocampal formation**, enabling scene construction and spatial imagination
- **replay and preplay mechanisms**, where the brain rehearses potential futures
- **integration areas** that combine disparate conceptual structures

These neural systems provide the substrate for structural M/A transitions.

They enable the mind to explore possibility space without immediate external input.

18.4 Imagination as Structural Generalization

Imagination is not random.
It is **structured exploration** governed by conceptual patterns.

The mind:

- recombines existing concepts
- interpolates between known patterns
- extrapolates trends
- fills in gaps
- rearranges components
- considers alternatives
- constructs analogies

This recombination follows modal rules:

- hypothesis (M/A)
- shaped by boundary (A/E)
- constrained by rails (M/M)
- guided by perceptual anchors (E/E)

Imagination is bounded creativity,
not unconstrained drift.

18.5 Hypothesis-Collapse and the Birth of Planning

Once the mind can explore possibilities internally,
it can evaluate them **before acting**.

This is the foundation of planning:

1. conceptual patterns (M/E)
2. freed into hypothetical variants (M/A)
3. tested through prediction and evaluation (E/M)
4. stabilized as strategies (A/M)
5. repeated into reliable practice (M/M)

Planning collapses imagination into action.

Hypothesis-collapse is the first step in that arc.

18.6 Creativity as Multi-Pattern Hypothesis-Collapse

Creativity emerges when:

- multiple conceptual patterns (M/E)
- collapse into a shared hypothesis space (M/A)
- forming new structures neither pattern alone could produce.

This explains:

- artistic intuition
- scientific innovation
- metaphor
- invention
- sudden insight ("Aha!" moments)

Insight is a multi-source M/A collapse
that resolves into a novel E/E understanding.

Creativity is collapse recombination.

18.7 Understanding Without Direct Experience

Hypothesis-collapse allows the mind to learn from:

- stories
- symbolism
- analogy
- instruction
- demonstration
- counterfactual reasoning

For example:

- A child can understand "fire is hot" before touching it.
- A scientist can understand a theoretical particle without observing it directly.
- A navigator can anticipate a route they have never traveled.
- A culture can imagine deities, spirits, or systemic forces.

Understanding through imagination is made possible by M/A.

Collapse does not always require direct perceptual evidence.
Internal structure can collapse into understanding.

18.8 Pathologies of Hypothesis-Collapse

When M/A becomes unstable or unconstrained, imagination can misfire:

Excessive M/A

- delusion
- runaway fantasy
- conspiratorial cognition
- rigid belief detached from evidence

Insufficient M/A

- lack of creativity
- difficulty imagining alternatives
- cognitive rigidity

Boundary mismatch (A/E drift)

- fear-based ideation
- intrusive thoughts
- catastrophic scenarios

Healthy imagination requires the right balance of:

- conceptual structure (M/E)
- internal boundaries (A/E)
- exploratory generativity (E/A)
- rail constraints (M/M)

Too much or too little of any leads to collapse dysfunction.

18.9 Hypothesis-Collapse Prepares the Ground for Prediction

Once a hypothesis exists,
the mind must **test** it.

This is expectation-collapse:

M/A → E/M

Prediction is hypothesis meeting boundary.

Understanding moves from:

- the possible
 to
- the probable
 to
- the actual.

Hypothesis-collapse opens possibility.
Expectation-collapse evaluates it.

Chapter 19 explores this transformation —
the basis of learning, inference, and scientific reasoning.

CHAPTER 19 — Expectation, Experimentation, and Learning

How the Mind Tests Its Possibilities Against the World

Hypothesis-collapse gives the mind a landscape of possibilities —
internal structures that can vary, combine, or be projected into imagined futures.
But possibilities alone do not produce understanding.
They must be tested.

Testing requires **expectation**:
a directed, structured anticipation of what should occur under given conditions.

In QC terms, this is:

M/A → E/M
possibility → prediction
hypothesis → experiment

Expectation-collapse is how the mind brings internal structure into contact with the external world.
It is the basis of learning.

19.1 What Expectation-Collapse Is

Expectation-collapse occurs when:

- the mind selects a specific possibility (from M/A),

- applies it to a particular boundary or situation (A/E),
- and anticipates a concrete outcome.

This anticipation is not a guess.
It is a structured hypothesis brought into alignment with sensory and contextual constraints.

Expectation-collapse is the moment when:

- an imagined outcome becomes a prediction,
- a prediction becomes a test,
- and a test produces new structure.

Learning begins with expectation.

19.2 Prediction as Structured Contact With Reality

Prediction is not foresight in the mystical sense.
It is the structural alignment between:

- patterns learned from the past (M/E),
- possibilities generated internally (M/A),
- and the constraints of the present (A/E).

When this alignment stabilizes,
the system makes a **prediction** (E/M) —
a collapse of internal possibility into anticipated action or outcome.

Examples include:

- anticipating where an object will fall,
- predicting the next word in a sentence,
- expecting a caregiver to respond to distress,
- imagining the consequences of a choice.

Expectation-collapse roots imagination in reality.

19.3 Expectation-Collapse Drives Learning

When a prediction is tested by the world,
one of two things happens:

1. Confirmation

The outcome matches the expectation.
The system strengthens the collapses that produced it.

2. Prediction Error

The outcome diverges from expectation.
The system updates its patterns, boundaries, or possibilities.

Learning is the recursive cycling of:

1. pattern (M/E)
2. possibility (M/A)
3. prediction (E/M)
4. outcome (E/E)
5. pattern revised (M/E again)

Every collapse in cognition feeds the next.

133

19.4 Neural Mechanisms of Expectation-Collapse

Neuroscience captures expectation in multiple forms:

- **predictive coding:**
  the brain continuously anticipates sensory input and minimizes prediction error
- **dopamine signaling:**
  reward prediction error shaping future expectations
- **motor forward models:**
  predicting the sensory consequences of movement
- **prefrontal modulation:**
  applying rules and contexts to guide expectations

These represent the physiological realization of QC's structural transformation.

Expectation-collapse is neurally embodied.

19.5 The Role of Boundary (A/E) in Prediction

Expectation requires boundaries:

- sensory context
- past experience
- environmental constraints
- social cues
- bodily conditions

- emotional states

Without boundary, predictions drift.
Without constraint, expectation becomes fantasy.
Without grounding, imagination loses traction.

Expectation-collapse is the moment
hypothesis meets the real.

19.6 Experimentation as Cognitive E/M

Experimentation is simply expectation-collapse made explicit.

To experiment is to:

- form a hypothesis (M/A)
- identify conditions (A/E)
- make a prediction (E/M)
- observe the outcome (E/E)
- update the underlying structure (M/E)

This is how science formalizes the natural learning loop.

Every scientist uses the mind's inherent modal machinery, but on the world's terms.

19.7 When Expectation Fails

Prediction failure is not a mistake.
It is a collapse with meaning.

Errors drive learning:

- The mind revises boundaries (A/E).
- It reshapes patterns (M/E).
- It modifies possibilities (M/A).
- It alters future predictions (E/M).
- It strengthens or weakens rails (M/M).

Prediction failure is the engine of adaptation.

Collapse not only creates structure —
it reshapes it.

19.8 Learning as Nested Collapse

Learning is not accumulation.
It is nested collapse:

1. **Object-collapse:**
   perception stabilizes sensory input.
2. **Law-collapse:**
   patterns emerge across experiences.
3. **Hypothesis-collapse:**
   new possibilities appear.
4. **Expectation-collapse:**
   possibilities are tested.
5. **Room-collapse:**
   stable contexts form for structured behavior.
6. **Rail-collapse:**
   persistent skills and habits emerge.

Every level of learning is a collapse within collapses.

The mind restructures itself continuously.

19.9 The Emergence of Skill and Expertise

Skill emerges when:

- repeated predictions succeed,
- contexts become familiar,
- patterns become reliable,
- procedural boundaries stabilize.

This is rail-collapse in cognition:

A/M → M/M

Skill is the internalization of successful collapses.
Expertise is collapse perfected.

The novice relies on prediction;
the expert relies on rails.

Expectation-collapse is the bridge between them.

19.10 Transition to Constructed Contexts (Room-Collapse)

Once predictions stabilize,

- the mind begins to build internal contexts
- to support reliable thinking and action.

This is room-collapse in cognition (E/M → A/M):

- schemas
- task sets
- interpretive frameworks
- cognitive scaffolding
- conceptual "environments"

Cognition now begins to organize itself.

The next chapter, Chapter 20, explores this structural transformation.

CHAPTER 20 — Constructed Contexts and Habit Formation

How the Mind Builds Internal Boundaries and Stabilizes Them

Expectation-collapse gives the mind the ability to test ideas against the world.
But testing alone is not enough to create coherence, memory, stable learning, or identity.
For understanding to become durable, the mind must learn to **construct internal boundaries** — contexts that stabilize meaning and shape future collapses.

This is the cognitive expression of:

Room-collapse (E/M → A/M)
and
Rail-collapse (A/M → M/M)

Room-collapse allows cognition to build its own interpretive space.
Rail-collapse gives that space permanence.

Together, they form the architecture of thought.

20.1 Room-Collapse: The Mind Builds Its Own Boundaries

In room-collapse, the mind reorganizes itself around stable expectations.

A context becomes explicit — not imposed by the environment,
but internally constructed.

Examples include:

- cognitive schemas
- working memory sets
- conceptual frameworks
- narrative structures
- problem-solving heuristics
- attentional biases
- task-specific mental "rooms"

Room-collapse is how the mind shapes the stage on which further collapses occur.

It is internal boundary formation.

20.2 Why Room-Collapse Is Necessary for Coherent Thought

Without room-collapse:

- learning remains episodic
- memory lacks organization
- perception is fragmented
- predictions lack stability
- identity cannot form
- behavior is inconsistent
- reasoning becomes reactive instead of generative

Room-collapse provides:

- coherence
- structure
- context
- persistence
- interpretive depth
- the ability to hold a problem in mind

It is the foundation of cognitive continuity.

20.3 Neural Correlates of Constructed Contexts

Room-collapse corresponds to neural processes including:

- **prefrontal cortex activation:** creating task sets and cognitive control
- **parietal association networks:** integrating spatial and conceptual context
- **working memory loops:** maintaining ideas over time
- **schema networks:** long-term stores of organized knowledge
- **attention networks:** selecting which information enters context

These systems allow the brain to build its own constraints for thought.

20.4 How the Mind Uses Constructed Contexts

Room-collapse influences cognition by providing:

1. Interpretive Boundaries

What counts as relevant or irrelevant.

2. Behavioral Expectations

"Given this situation, this set of actions makes sense."

3. Cognitive Shortcuts

Predictive frames that reduce processing load.

4. Memory Organization

Information slots into structured conceptual spaces.

5. Role-Based Thinking

"I am learning," "I am solving," "I am imagining."

Context gives meaning to experience.
Room-collapse builds meaning's scaffolding.

20.5 Constructed Contexts Enable Complex Behavior

Humans rely heavily on constructed contexts:

- language frames conversation
- stories frame interpretation

- roles frame interaction
- institutions frame social behavior
- scientific paradigms frame inquiry
- cultural narratives frame identity

In individuals, these appear as:

- interpretive schemas
- internalized norms
- rule-based reasoning
- cognitive strategies
- symbolic and conceptual rooms

Room-collapse transforms experience from a stream into a structured environment.

20.6 Rail-Collapse: When Context Becomes Habit

Some constructed contexts stabilize so strongly that they become **default modes of behavior**.

This is rail-collapse:

A/M → M/M
constructed context → stable habit

Examples include:

- established motor skills
- linguistic fluency
- practiced reasoning strategies
- internalized moral frameworks

- persistent emotional tendencies
- deeply learned cognitive patterns

Rails save energy, increase coherence,
and provide stable identity anchors.

Habit is structured collapse made automatic.

20.7 Why Rails Are Double-Edged

Rails are indispensable —
but they are also limiting.

Rails enable:

- fast decision-making
- reduced cognitive effort
- strong identity
- reliable behavior
- skill mastery
- mental efficiency

Rails limit:

- flexibility
- creativity if over-rigid
- ability to reinterpret situations
- openness to novelty
- adaptation under stress

Rails are commitments.
They make the mind consistent,
but they also make it predictable.

Rail-collapse gives us identity,
and identity brings both stability and constraint.

20.8 Rail-Thawing: How the Mind Regains Flexibility

Rails can be softened or dissolved when:

- context shifts
- prediction errors accumulate
- emotional or social pressures demand reevaluation
- reflective cognition intervenes
- new learning destabilizes old assumptions
- trauma forces structural reorganization

Thawing is not failure.
It is modal flexibility.

The mind constantly negotiates between rails and possibility.

Healthy cognition warms and cools rails as needed.

20.9 Constructed Contexts Are the Bridge to Culture

Individual room-collapse and rail-collapse
scale up into cultural forms:

- norms

- rituals
- languages
- institutions
- roles
- shared values

Culture is **cognition distributed across many minds**, with room-collapse and rail-collapse operating collectively.

The next part of the book rises from the individual mind to the shared structures of culture.

Chapter 21 begins this ascent by showing how identity formation leads into cultural roles.

CHAPTER 21 — Cognitive Rail-Collapse and the Formation of Identity

How Patterns Become Habits, Habits Become Identity, and Identity Becomes Architecture

Constructed contexts (A/M) give cognition the ability to organize experience.
But true stability — the kind that makes thinking efficient, memory reliable, and behavior coherent — requires a deeper form of collapse:
the stabilization of these constructed contexts into **rails**, the mind's enduring structures.

This is **rail-collapse** in cognition:

A/M → M/M
context → stable mode
habit → identity

Rail-collapse is how the mind becomes itself.

21.1 What Cognitive Rails Are

A cognitive rail is a stable behavior or thought pattern that:

- activates automatically under certain conditions
- requires little processing effort
- organizes interpretation
- guides decisions

- remains consistent across time
- persists even under moderate disruption

Rails include:

- habits
- motor skills
- learned procedures
- semantic networks
- established reasoning styles
- emotional defaults
- identity structures

Rails are cognition's long-term architecture.

21.2 How Rails Form: From Repetition to Stability

Rails begin with:

- repeated expectations (E/M)
- functioning reliably across contexts
- constructed into mental rooms (A/M)

When these rooms are activated often enough, the boundaries and processes that define them:

- become automatic
- require no reconstruction
- stabilize into persistent patterns

This is rail-collapse.

It is the QC equivalent of:

- synaptic consolidation
- procedural memory formation
- automatization
- personality consolidation
- long-term learning

Rail-collapse makes cognition fast and coherent.

21.3 Rails as Cognitive Inheritance

Rails store:

- what worked
- what mattered
- what succeeded
- what kept the organism alive
- what repeated enough to be trusted

Rails are the inheritance of experience.

Just as genes store biological history,
rails store cognitive history.

Everything the mind has collapsed successfully accumulates as rail structure.

Rails make understanding cumulative.

21.4 Identity as a System of Rails

Identity is not a single structure.
It is a **configuration of many rails**:

- cognitive rails (thinking styles)
- procedural rails (skills)
- emotional rails (affective tendencies)
- narrative rails (self-stories)
- social rails (roles internalized from culture)

Identity emerges when these rails:

- stabilize,
- coordinate,
- and become mutually reinforcing.

Identity = coherent M/M across many domains.

It is not static.
It is stable.

21.5 The Strength and Fragility of Rails

Rails bring strength:

- stability in uncertainty
- speed of action
- predictability of behavior
- conservation of cognitive resources
- deep expertise
- a consistent sense of self

But they also bring fragility:

- rigidity
- resistance to change
- difficulty adapting
- susceptibility to distortion
- vulnerability under context shift
- entrenchment of outdated assumptions

Rails are commitments.

They make understanding enduring,
but also directional.

21.6 Rails and Cognitive Style

Different individuals have different rails:

- some emphasize analytic rails
- some emotional rails
- some sensory or spatial rails
- some social interpretation rails
- some linguistic or symbolic rails

Rails determine:

- how we interpret the world
- how we solve problems
- where we excel
- where we struggle
- what we ignore
- what we foreground

Personality is the surface expression of rail architecture.

21.7 Rail-Drift: When Identity Evolves

Rails can drift when:

- new experiences accumulate
- persistent prediction errors emerge
- a major life transition shifts context
- trauma destabilizes established patterns
- therapy or reflection reconfigures boundaries
- culture exerts strong new pressures

Drift does not mean collapse.
It means **reorganization of stability**.

Identity grows when rails shift in ways that bring coherence.
Identity fractures when rails shift without alignment.

21.8 Rail-Thawing: When Change Becomes Necessary

Rails can and must thaw under certain conditions:

- new demands
- new environments
- new relationships
- new challenges
- internal mismatch
- suffering caused by rigid patterns

The mind can soften rails through:

- reflection
- practice
- new experiences
- emotional insight
- cognitive restructuring
- intentional intervention

Rail-thawing is how transformation happens.

Rigid rails break.
Flexible rails adapt.

21.9 Rails as the Bridge Between Cognition and Culture

Rails do not form only within the individual.

They are influenced by:

- cultural norms
- language
- shared stories
- roles
- traditions
- social institutions

These external rails shape:

- moral reasoning
- identity categories
- emotional expression

- attention biases
- conceptual frameworks

Cognition inherits rail structure from culture.
Culture inherits rail structure from cognition.

This mutual inheritance prepares the ground
for collective collapse at the scale of societies.

Chapter 22 begins the cultural arc.

Part IV Culture

CHAPTER 22 — Shared Field-Collapse: Collective Attention

How Minds Begin to Perceive Together

Cognition does not remain solitary.
When multiple minds interact, their individual collapses begin to influence one another.
Before a culture can form roles, norms, or shared meanings, it must first achieve the simplest kind of collective structure:

shared attention.

Shared attention is the cultural expression of **field-collapse**:

$A/A \to E/A + A/E$,
not inside a single mind,
but across many minds simultaneously.

This chapter describes the first structural step in cultural emergence:
how individual experiential fields collapse into a coordinated social field.

22.1 The Undivided Social Field (A/A Across Minds)

Before any cultural coherence emerges,
a group exists in a state analogous to cognitive A/A:

- individuals attend to separate stimuli
- no shared focus exists
- perceptions are uncoordinated
- relevance is fragmented
- social meaning has not yet formed

This undifferentiated social field is not disorganized;
it is simply unaligned.

There is no "we" yet.
Only many individual streams of experience.

22.2 The First Shared Focus (E/A Emerges Collectively)

Shared field-collapse begins when many individuals direct their attention toward the same stimulus:

- a predator
- a food source
- a sound
- a movement
- a member of the group
- an environmental cue
- an emotionally salient display

This alignment of attentional process (E/A)
does not require intention or language.

It emerges from:

- perceptual salience
- urgency
- attraction
- imitation
- contagion of arousal
- social coordination tendencies

Shared focus is the root of all collective structure.

22.3 Shared Boundaries (A/E Across Minds)

For a group to share attention,
its members must agree — implicitly —
on what counts as:

- a boundary
- a threat
- an opportunity
- an object of importance

Shared boundaries include:

- direction from which danger comes
- location of food or resources
- social signals of dominance or submission
- emotional expressions
- environmental landmarks

This is **collective A/E**,
the emergence of socially synchronized constraint.

The group begins to "see" the same edges.

22.4 Why Shared Attention Is Necessary for Culture

Culture is collapse at scale.
It requires:

- shared percepts
- shared objects
- shared patterns
- shared contexts
- shared meaning
- shared behavior

None of this is possible
until a group undergoes shared field-collapse.

No roles can form if a group cannot perceive the same kinds of social distinctions.
No norms can stabilize if a group cannot attend to the same disruptions.
No rituals can emerge if a group cannot synchronize behavior.

Shared attention is the foundation of cultural structure.

22.5 Evolutionary Evidence of Collective Field-Collapse

Animals across many lineages exhibit shared field-collapse:

Primates

- coordinated vigilance
- joint gaze
- attention to emotional displays
- mimicry and social referencing

Birds and Fish

- flocking and schooling
- synchronized direction shifts
- collective scanning for danger

Mammalian Herds

- coordinated movement
- group orientation toward predators
- synchronized feeding or resting patterns

Humans

- joint attention in infants
- shared focus during rituals, storytelling, music, play
- collective attention in classrooms, ceremonies, protests

Evolution built social field-collapse
long before culture emerged.

22.6 Collective E/E: The First Social Objects

Once a group attends together (E/A)
and shares boundaries (A/E),
object-collapse becomes possible at the group scale.

The group begins to perceive:

- shared roles
- shared signals
- shared threats
- shared opportunities
- shared meanings in movement or sound

These are the first **social E/E** —
proto-cultural objects.

Not internal to one mind,
but distributed across many minds.

This is the beginning of shared reality.

22.7 Cultural Field-Collapse as the Root of Social Cognition

Field-collapse across minds enables:

- joint attention
- coordinated action
- shared inference
- collective emotion
- imitation and learning
- social contagion
- synchronization of behavior

Humans do this with extraordinary sensitivity, but the structure appears in simpler forms across species.

Social cognition is the internalization of collective field-collapse.

It is the modal bridge between individual and group.

22.8 Why Cultural Collapse Mirrors Cognitive Collapse

The parallel is precise:

Cognition:
A/A → undifferentiated sensation
E/A + A/E → focus + contrast
E/E → percepts

Culture:
A/A → undifferentiated social field
E/A + A/E → shared attention + shared boundaries
E/E → social objects

Cultural collapse is cognition scaled across many minds.

Culture is a distributed mode of understanding.

22.9 Transition to Social Object-Collapse

When shared attention stabilizes,
and groups repeatedly attend to the same boundaries and events,
their interactions begin to collapse into:

- stable social objects
- roles
- signals
- shared responses
- coordinated meanings

This is **social object-collapse**,
the structural foundation of norms, roles, and early cultural meaning.

Chapter 23 explores this next transformation.

CHAPTER 23 — Social Object-Collapse

How Groups Form Shared "Things" That No Individual Alone Could Create

Once a group achieves shared attention and shared boundaries,
it can undergo the next major cultural transformation:
the formation of **social objects** — stable, collectively recognized structures that shape behavior.

These social objects are not physical things.
They are **shared cognitive constructions**,
reinforced across many minds through repeated interaction.

This is **social object-collapse**,
the distributed equivalent of percept formation:

$E/A + A/E \rightarrow E/E$
(**shared focus + shared boundary $\rightarrow$ shared object**)

Shared objects become the raw materials of culture.

23.1 What Social Objects Are

A social object is a pattern that:

- many individuals perceive together,
- persists through repeated interactions,
- influences behavior,
- and becomes recognizable across contexts.

Examples include:

- a leader
- a threat
- a safe member
- a mating display
- an alliance
- a gesture with meaning
- a forbidden action
- a trusted partner
- a place of significance

These are not conceptual abstractions.
They are **collectively stabilized percepts**.

They arise through distributed object-collapse.

23.2 Why Social Objects Are Necessary for Culture

Without social objects, groups cannot develop:

- coordination
- stable roles
- norms
- rituals
- shared expectations
- collective memory
- social identity
- cooperation
- institutions

A culture is impossible without objects it collectively recognizes.

Social object-collapse is the birth of shared meaning.

23.3 How Social Object-Collapse Works

Social object-collapse requires:

1. **Shared attention (E/A):**
   the group focuses on the same event.
2. **Shared boundaries (A/E):**
   the group interprets cues similarly.
3. **Repetition:**
   the same interactions happen again and again.
4. **Convergence:**
   individuals treat the event or behavior as "the same thing."
5. **Stability:**
   the social object persists even when contexts vary.

When these conditions align,
the group collapses repeated interactions
into a **collective object**.

The object exists between minds,
not inside any single one.

23.4 Examples Across Species

Primates

- dominant individuals
- grooming relationships
- threat displays
- known rivals
- safe companions

Birds

- courtship dances
- alarm call meanings
- territorial boundaries

Mammals

- signals of submission
- shared migration routes
- coordinated hunting roles

Humans

- greetings
- taboos
- ritual gestures
- roles such as "teacher," "child," "elder," "stranger"

These are not merely learned behaviors.
They are culturally collapsed objects.

23.5 Social Objects Are Not Inside the Brain

A social object:

- is reinforced across many minds
- exists only through group dynamics
- is maintained by collective behavior
- is distributed rather than internal

It is a **shared perceptual construction**, relying on:

- alignment
- imitation
- reinforcement
- memory
- expectation

The object is real because the group treats it as real.

Culture is held in the space between individuals.

23.6 Social Object-Collapse and Meaning

Once social objects exist, groups begin to:

- respond reliably to the same signals
- share emotional reactions
- coordinate actions
- form stable expectations
- transmit meaning across generations

Shared objects allow minds to synchronize meaning.

This is the foundation of:

- teaching
- cooperation
- role adoption
- social cohesion
- group decision-making

Meaning becomes collective.

23.7 Social Objects and the Origins of Norms

Stable social objects naturally lead to:

- repeated behavioral patterns
- emergent expectations
- proto-norms
- implicit agreements
- coordinated responses
- regularized social interactions

These patterns undergo **cultural law-collapse** (E/E → M/E),
creating the first enduring norms.

Social objects → social patterns → cultural laws.

Every culture builds itself through this sequence.

23.8 Social Object-Collapse Prepares the Ground for Tradition

As social objects stabilize,
they accumulate history:

- who did what
- how interactions resolved
- how behavior affected outcomes
- how groups responded
- what was rewarded or punished

This history becomes:

- stable norms
- proto-traditions
- early ritual structures

Social object-collapse contains the seeds of culture's future rails.

23.9 Transition to Cultural Law-Collapse

Once groups consistently perceive the same social objects, they begin to detect *patterns* across them:

- repeated interactions
- recurring sequences
- typical behaviors
- stable social responses
- reliable outcomes

This leads to:

E/E → M/E
social objects → cultural regularities

Cultural law-collapse is the next step —
how shared objects become shared norms.

Chapter 24 explores this transition.

CHAPTER 24 — Cultural Law-Collapse

How Shared Social Objects Become Norms, Traditions, and Stable Patterns

Once a group forms stable social objects through repeated collective interactions,
it becomes capable of detecting **patterns** across those objects.
These patterns arise not from individual minds but from the **collective behavior of many minds over time**.

When social objects recur in similar contexts with similar consequences,
they undergo:

E/E → M/E
social object → cultural regularity
repeated event → norm

This is **cultural law-collapse** —
the emergence of norms, conventions, and early traditions.

24.1 Why Shared Objects Naturally Produce Shared Patterns

When social objects stabilize, groups begin to:

- anticipate how others will respond
- synchronize behaviors
- form expectations about interactions

- repeat successful sequences
- avoid harmful or destabilizing actions
- recognize stable social outcomes

Across repeated cycles, patterns condense.

Shared object → shared response → stable expectation → cultural law.

Cultural law-collapse transforms social behavior from improvisational to structured.

24.2 Cultural Law-Collapse Across Species

This process is not uniquely human.
Many species exhibit stable patterns derived from shared social objects:

Primates

- grooming hierarchies
- alliance structures
- maternal care routines
- conflict-resolution sequences

Elephants

- mourning behaviors
- greeting ceremonies
- group caretaking patterns

Birds

- coordinated mating dances
- long-distance migratory routes
- shared alarm-call meanings

Cetaceans

- hunting strategies
- pod-specific vocal patterns
- social play rituals

These are cultural laws in their earliest forms: repeated collapses producing stable behavioral patterns.

24.3 Cultural Regularities Are Emergent, Not Imposed

Cultural laws form organically.
They arise from:

- repetition
- constraint
- shared boundaries
- shared attention
- successful coordination
- group-level reinforcement

No central authority is required.
Norms appear because collapse naturally selects stable solutions.

In this sense:

Norms are the patterns the group discovers about how to remain coherent.

Cultural law-collapse is the collective learning of a social system.

24.4 The Structure of Cultural Law-Collapse

Cultural law-collapse requires:

1. Stable social objects (E/E)

Roles, signals, places, relationships.

2. Repetition

The same kinds of interactions recur.

3. Boundary conditions (A/E)

Group-wide constraints that affect behavior.

4. Convergence

The group tends toward similar responses.

5. Persistence

The pattern survives across many cycles of interaction.

When these conditions align,
the group's behavior collapses into stable cultural patterns.

24.5 Cultural Laws Are Not Perfect — They Are Robust

Cultural laws:

- tolerate variation
- adapt to context
- drift when conditions change
- differ between groups
- are shaped by environment
- can be flexible or rigid
- may persist long after original conditions vanish

They are robust rather than mathematical.

Unlike physical laws, cultural laws:

- can be bent
- interpreted
- re-negotiated
- resisted
- reinvented

But they remain sufficiently stable
to guide group cohesion and behavior.

24.6 Cultural Laws as Templates for Coordination

Cultural regularities serve as templates for:

- conflict resolution
- resource sharing
- mating and kinship
- social hierarchy
- cooperation
- communication
- group movement
- ritual and ceremony
- moral judgment (in humans)

These templates reduce:

- uncertainty
- cognitive load
- social conflict
- coordination costs

Cultural law-collapse is efficiency at the group level.

24.7 From Patterns to Traditions: M/E in Cultural Memory

As patterns repeat across generations,
they form early traditions:

- ways of greeting
- feeding order

- care patterns
- emblems or displays
- shared places of importance
- routines marking transitions or seasons

Traditions are **cultural M/E** carried forward in memory —
sometimes embodied in behavior,
sometimes reinforced through imitation,
sometimes scaffolded by environment.

Tradition is the beginning of cultural continuity.

24.8 The Role of Emotion in Cultural Law-Collapse

Emotion is one of the mechanisms that binds patterns into norms:

- shared fear
- shared joy
- shared mourning
- shared bonding
- shared anger

Emotion amplifies:

- salience
- memory
- synchrony
- reinforcement

Patterns repeated with emotional resonance
collapse more strongly into cultural regularities.

Emotion is the adhesive of early culture.

24.9 Cultural Law-Collapse Prepares for Cultural Hypothesis and Experimentation

Once patterns are stable,
groups can:

- elaborate rituals
- explore symbolic variation
- develop mythic narratives
- invent new tools or strategies
- experiment with norms
- test alternative forms of coordination

This is the cultural equivalent of cognitive hypothesis- and expectation-collapse:

$M/E \rightarrow M/A \rightarrow E/M$

Culture begins to **imagine itself**.

The next chapter explores this generative expansion.

Chapter 25 examines how possibility and experimentation arise in cultural systems.

CHAPTER 25 — Cultural Hypothesis- and Expectation-Collapse

How Groups Imagine Possibilities and Test Them Through Collective Action

Once cultural patterns stabilize through cultural law-collapse,
a group gains a new capacity that parallels the rise of imagination in the individual mind:
the ability to **explore shared possibilities**.

This involves two linked transformations:

M/E → M/A
patterns → shared possibilities
regularities → collective imagination

and

M/A → E/M
possibilities → coordinated predictions
hypotheses → tests in behavior

These are **cultural hypothesis-collapse** and **cultural expectation-collapse**,
the engines of collective creativity, adaptation, ritual, and innovation.

25.1 Cultural Hypothesis-Collapse: Possibility Shared Across Minds

When a cultural pattern is sufficiently stable,
the group can begin to **vary** it —
to imagine different forms it could take.

Examples include:

- elaborating a ritual gesture into a ceremony
- extending a cooperative behavior into coordinated strategy
- interpreting a symbol in new ways
- generating stories about origins or meaning
- experimenting with new social practices
- playing or improvising within a familiar structure

This is **M/A at the group level** —
patterns freed from their original contexts
and explored collectively.

Cultural imagination is the earliest form of symbolic behavior.

25.2 Collective Possibility: How Groups "Imagine"

Cultural hypothesis-collapse does not require:

- language,
- consciousness,
- abstract thought,

- or explicit reasoning.

It can arise through:

- shared play
- coordinated movement
- repeated improvisation
- emotional synchrony
- social learning
- shared curiosity
- imitation of innovations

Groups begin to explore variations of behaviors that once seemed fixed.

Possibility reveals itself through interaction.

25.3 Myth as Cultural Hypothesis Space

Myth emerges as:

- shared narrative possibility
- symbolic recombination
- pattern extended into story
- collective imagination given form

Myth is not primitive error;
it is the cultural expression of **hypothesis-collapse**.

Myths:

- unify patterns of meaning

- expand interpretive space
- encode shared emotional experience
- provide collective expectations
- structure possible futures

Myth is how early cultures explore what could be true.

25.4 Cultural Expectation-Collapse: Testing Shared Possibilities

Once a possibility is collectively imagined,
the group can **test** it through coordinated action.

This is **cultural E/M**:

- a ritual variation is tried
- a new signaling pattern is attempted
- a hunting strategy is coordinated
- a social arrangement is experimented with
- a narrative is used to guide behavior
- a symbolic marker is introduced

Some tests succeed.
Some fail.
Outcomes shape the next collapse.

Cultural expectation-collapse is collective experimentation.

25.5 When Possibilities Become Predictions

Cultural predictions are implicit but real:

- "If we dance this way, the group will synchronize."
- "If we display this symbol, others will respond accordingly."
- "If we organize in this formation, the hunt will succeed."
- "If we perform this ritual, emotional cohesion will rise."
- "If we cooperate in this arrangement, conflict will decrease."

These are not conscious hypotheses.
They are **structural expectations** applied through behavior.

Cultural E/M is the testing of shared imagination.

25.6 Experimentation Is Not Always Deliberate

Cultural experiments arise from:

- imitation
- drift
- accidental discovery
- environmental pressure
- social conflict and resolution
- error and recovery
- spontaneous coordination
- play and improvisation

Even without explicit intention,
cultural systems continually test variations.

Expectation-collapse channels variation into outcomes
that the group can incorporate or reject.

25.7 Cultural Success and Failure as Collapse Outcomes

Successful collapses become:

- new traditions
- stronger roles
- refined rituals
- innovations
- strengthened cooperation
- enhanced group survival

Failed collapses:

- vanish
- are suppressed
- are avoided in the future
- or become cautionary tales

Culture "learns" structurally,
not conceptually.

Expectation-collapse is the filter
through which cultural novelty becomes cultural stability.

25.8 Hypothesis, Expectation, and the Roots of Innovation

Innovation emerges from:

1. a stable pattern (M/E)
2. variation or recombination (M/A)
3. collective testing (E/M)
4. stabilization (A/M)
5. repetition until persistent (M/M)

This sequence produces:

- new tools
- new strategies
- new rituals
- new narratives
- new social arrangements
- new technologies
- new institutions

Cultural innovation is the modal recursion of collapse.

25.9 Cultural Hypothesis- and Expectation-Collapse Prepare for Institutions

Once a group reliably tests its imagined structures,
it can begin to **construct** the boundaries that stabilize them:

- roles formalize

- rituals ritualize
- norms codify
- conventions solidify
- meanings anchor
- expectations harden
- cooperative structures become institutions

This is cultural **room-collapse**
(E/M → A/M across many minds).

Chapter 26 explores how groups construct the "rooms" in which collective meaning and social order emerge.

CHAPTER 26 — Constructed Cultural Boundaries

How Groups Build Rituals, Institutions, and Shared Constraints

Once a culture can imagine possibilities (M/A)
and test them through coordinated action (E/M),
it gains a new and transformative capacity:
the ability to **construct the boundaries** that stabilize its shared behavior.

This is cultural **room-collapse**:

E/M → A/M
tested possibility → constructed context

and eventually, cultural **rail-collapse**:

A/M → M/M
constructed context → enduring tradition or institution

These collapses create the scaffolding of culture:
rituals, norms, institutions, social categories,
shared values, and collective identities.

26.1 What Cultural Room-Collapse Is

Cultural room-collapse occurs when the group:

- develops a shared expectation (E/M),

- recognizes that this expectation reliably produces stability,
- and constructs a **shared boundary or structure** to support it (A/M).

These constructed cultural boundaries can be:

- ritual sequences
- group roles
- kinship systems
- shared symbolic markers
- gathering places
- social rules
- division of labor
- proto-institutions
- shared calendars or cycles
- agreed-upon procedures

Room-collapse is not imposed from above.
It emerges from collective behavior.

The group builds the constraints
that make its own coherence possible.

26.2 Ritual as Constructed Boundary

Rituals emerge when:

- a repeated behavior
- reliably produces emotional cohesion or coordination (E/M)
- and the group formalizes the sequence (A/M)

Examples include:

- greeting rituals
- courting dances
- communal meals
- synchronized movement or chanting
- mourning practices
- initiation rites
- seasonal observances

Rituals create **rooms in time**:
shared moments with structured boundaries
that shape collective experience.

26.3 Roles as Cultural Rooms

Roles are **rooms in social position**.

They arise when:

- individuals regularly occupy certain functions,
- expectations become stable,
- and the group constructs boundaries around those functions.

Examples include:

- leader
- elder
- healer
- caretaker
- scout
- mediator

- storyteller
- outsider
- initiate

Roles allow groups to coordinate behavior
by distributing responsibilities across a social map.

Room-collapse organizes social structure.

26.4 Norms as Constructed Constraints

A norm is a **shared boundary** that emerges when:

- a pattern stabilizes
- a group recognizes its stability
- and behavior is organized around it

Norms include:

- reciprocity
- fairness expectations
- conflict-avoidance behaviors
- mating rules
- food taboos
- territory or resource rules
- forms of politeness or respect

Norms reduce uncertainty
and make social collapse predictable.

They are cultural A/E constructed from repeated E/M.

26.5 Institutions: The Deep Rooms of Culture

Institutions arise when constructed boundaries:

- persist across generations
- become self-reinforcing
- accumulate symbolic significance
- formalize roles and norms
- coordinate large numbers of individuals
- maintain stability even as members change

Institutions include:

- kinship systems
- religious structures
- codes of conduct
- legal frameworks
- governance patterns
- economic arrangements
- educational systems

Institutions are cultural A/M at scale.

They are the primary rooms of civilization.

26.6 Cultural Rail-Collapse: When Constructed Boundaries Become Tradition

Once a constructed boundary becomes deeply stable
and self-reinforcing across time,
it undergoes cultural **rail-collapse (A/M → M/M)**.

Rails in culture include:

- rituals that survive centuries
- social roles that become archetypes
- kinship structures that define identity
- symbolic systems that endure (language, art, religion)
- institutions that persist across generations
- moral frameworks internalized by individuals

Cultural rails are patterns so deeply constructed
that they appear "natural."

They form the inherited architecture of social life.

26.7 The Benefits and Dangers of Cultural Rails

Benefits

Rails provide:

- stability
- continuity
- coordination
- collective identity
- long-term memory
- shared meaning
- reduced conflict

- efficient organization

Dangers

Rails also create:

- rigidity
- suppression of novelty
- exclusion or hierarchy
- difficulty adapting to change
- inertia
- institutional drift
- ossification of outdated structures

Rails enable thriving but can hinder transformation.

Culture evolves when rails are strong enough to support life but flexible enough to allow restructuring.

26.8 How Cultural Rooms Become Cognitive Rooms

Constructed cultural boundaries become **internalized cognitive boundaries**.

For example:

- a cultural role becomes part of personal identity
- a moral rule becomes a cognitive rail
- a ritual becomes an emotional schema
- a linguistic category shapes thought
- institutional expectations guide interpretation
- symbolic frameworks shape reasoning

Cognition inherits cultural rooms
and then builds upon them internally.

The relationship is recursive.

Culture shapes minds;
minds shape culture.

26.9 Transition to Cultural Stability (Rail-Collapse)

With constructed rooms in place,
and patterns of behavior stabilized across generations,
culture becomes capable of a final transformation:

the creation of long-lasting rails —

- traditions
- languages
- canonical stories
- moral frameworks
- institutional structures
- symbolic systems

These rails give culture its recognizable shape
and allow it to endure beyond individuals.

Chapter 27 explores this stabilization:
cultural rail-collapse and how civilizations take form.

CHAPTER 27 — Cultural Rails: Tradition, Inertia, Innovation, and Drift

How Societies Stabilize Their Structures Across Generations

Once cultures construct their own boundaries through rituals, roles, norms, and early institutions,
they become capable of a deeper form of stability:
the long-term preservation of these boundaries across generations.

This is **cultural rail-collapse**:

A/M → M/M
constructed boundary → durable structure
institution → tradition

Rails stabilize culture the way habits stabilize cognition. They hold the shape of a society over time.

27.1 What Cultural Rails Are

A cultural rail is a deeply entrenched pattern that:

- persists across generations
- organizes social behavior
- structures meaning
- constrains interpretation
- supports cooperation

- resists disruption
- stores cultural memory

Rails can be:

- traditions
- institutions
- roles
- rituals
- moral norms
- languages
- foundational stories
- economic practices
- governance structures

Rails are the durable architecture of culture.

27.2 Rails Form When Rooms Become Self-Reinforcing

Cultural rails emerge when:

1. a constructed boundary (A/M)
2. works reliably enough
3. for long enough
4. that the group comes to depend on it
5. and transmits it across time

This persistence converts:

- rituals → traditions
- norms → moral frameworks
- roles → social categories

- strategies → institutions
- stories → symbolic canons

Rails become **collective memory encoded in structure**.

27.3 Why Rails Matter

Rails perform critical functions:

- **Coordination:** large groups can act together
- **Continuity:** meaning and behavior persist across generations
- **Predictability:** individuals know what to expect
- **Stability:** social structures remain coherent
- **Efficiency:** rules reduce cognitive and social complexity
- **Identity:** groups know who "we" are

Rails are what make societies
recognizable, navigable, and enduring.

27.4 The Double-Edged Nature of Rails

While rails support stability, they also constrain possibility.

Rails enable:

- skill transmission
- trust networks

- intergenerational coherence
- institutional depth
- shared values
- cumulative knowledge

Rails limit:

- novelty
- structural flexibility
- reinterpretation
- diversity of expression
- adaptability to change

Rails maintain culture,
but they also create inertia.

Every civilization must navigate this tension.

27.5 How Cultural Rails Drift

Rails are stable but not static.
They drift through:

- reinterpretation
- demographic change
- environmental shifts
- technological innovation
- cross-cultural contact
- gradual erosion
- generational negotiation

Drift alters:

- meaning
- practice
- boundaries
- expectations
- institutional structure

Drift can be subtle or dramatic, but it is always present.

Rails evolve.

27.6 When Rails Break

Rails can also fail:

- sudden shocks
- war, collapse, famine
- environmental catastrophe
- technological disruption
- loss of trust
- internal contradiction
- revolution
- institutional decay
- cultural fragmentation

When rails fail:

- norms destabilize
- roles loosen
- institutions weaken
- identity fractures
- cohesion dissolves

The collapse of rails is cultural system failure.

But it also opens the door
to renewal and reinvention.

27.7 The Interplay of Innovation and Tradition

Cultures thrive when they balance:

- **rails** for stability
- **rooms** for adaptation
- **expectation-collapse** for innovation
- **hypothesis-collapse** for imagination
- **law-collapse** for coherence

Too much rail → stagnation.
Too little rail → chaos.

Successful civilizations maintain a dynamic equilibrium between stability and possibility.

Rails that are too rigid eventually shatter.
Rails that flex can endure.

27.8 How Cultural Rails Become Cognitive Rails

Cultural rails become:

- identity categories
- moral frameworks
- linguistic distinctions
- emotional norms
- conceptual scaffolds
- relational patterns
- reasoning defaults

These rails are internalized by individuals
through childhood development
and repeated social interactions.

Cognition does not operate independently of culture.
It inherits rails from culture
and then adds its own.

Rails link individual minds
to collective meaning structures.

27.9 Rails as the Foundation of Civilization

Civilizations arise when rails accumulate across domains:

- language rails
- economic rails
- religious rails
- governance rails
- social hierarchy rails
- technological rails
- scientific rails
- symbolic and narrative rails

These systems of rails:

- encode a civilization's history
- provide stability across centuries
- enable large-scale cooperation
- shape cultural identity
- guide behavior without explicit instruction
- create the conditions for innovation and drift

Civilization is not built out of objects.
It is built out of collapses that have become rails.

27.10 Transition to Artificial Systems

Culture is the highest natural scale
at which collapse dynamics have unfolded on Earth.

But collapse also appears
in a new, non-biological substrate:
artificial systems.

Part VI explores how QC manifests in computation,
how collapse becomes observable in machines,
and how artificial rails illuminate the mechanics
of cognition, culture, and emergence.

Chapter 28 begins the AI arc.

Part V Artificial systems

CHAPTER 28 — Making Collapse Visible in Machines

How Artificial Systems Reveal the Mechanics of Modal Transformation

Physics, biology, cognition, and culture all display collapse dynamics,
but in these natural systems collapse is always partially hidden:

- physics obscures collapse beneath mathematical formalism,
- biology obscures collapse beneath biochemical complexity,
- cognition obscures collapse beneath subjective experience,
- culture obscures collapse beneath distributed interaction.

Artificial systems are different.

For the first time in the history of science,
collapse can be **observed directly**,
measured step-by-step,
and analyzed without ambiguity.

This chapter introduces why artificial systems are uniquely suited
to reveal the generative structure QC describes.

28.1 Why Collapse Is Hard to Observe in Natural Systems

Natural systems hide collapse because:

Complexity

Interactions occur across vast numbers of components.

Scale

Some collapse happens too quickly or too subtly to measure.

Integration

Collapse interacts with multiple levels simultaneously.

Embedding

Systems operate within environments that shape collapse invisibly.

Opacity

In cognitive and cultural systems, collapse is internal or distributed.

Artificial systems remove these barriers.

Collapse in machines:

- leaves traces,
- produces logs,

- creates measurable transitions,
- and can be tested under controlled conditions.

28.2 Artificial Systems Contain the Same Modal Architecture

Even though artificial systems are not alive or conscious, they are dynamic, constrained systems that process information.

As such, they naturally instantiate QC's modal roles:

E/A — generative activity

A/E — architectural constraints

E/E — stable output representations

M/E — patterns learned from data

M/A — internal generalizations

E/M — predictions

A/M — contexts created by prompting or task design

M/M — stable inference habits or "rails"

The substrate differs.
The structure does not.

Anything that transforms process into pattern under constraints
will exhibit modal dynamics.

28.3 Why Artificial Systems Make Collapse Visible

In artificial systems:

- activation flows can be tracked
- state transitions can be logged
- outputs can be measured precisely
- collapse timing can be plotted
- drift can be quantified
- rails can be detected
- room construction can be controlled
- boundary conditions can be manipulated directly

These capabilities give researchers a level of access that biology and cognition do not.

AI becomes a **microscope for studying collapse**.

28.4 E/A and A/E in Artificial Systems

E/A — generative activity

In language models: token-generation dynamics.
In reinforcement learning: exploratory policies.

A/E — boundaries

These include:

- architecture
- training data
- loss functions
- safety constraints
- prompts
- system rules

Process and boundary interact directly, with no hidden variables.

28.5 Object-Collapse in Machines (E/A → E/E)

Artificial object-collapse occurs when:

- a generative process stabilizes
- into a coherent internal representation or output

Examples:

- a stable embedding vector
- a consistent output interpretation
- convergence to a predicted answer
- a policy stabilizing after exploration

E/E events are easy to detect in machines because collapse is accompanied by stable activation patterns.

Machines reveal cognitive object-collapse with unmatched clarity.

28.6 Pattern Formation as Machine Learning (E/E → M/E)

Machine learning is essentially:

repeated object-collapses → stable patterns

Models extract:

- regularities
- invariances
- associations
- structural relationships

Training is law-collapse in a digital substrate.

Models "learn" only in the structural QC sense —
not through meaning or experience
but through pattern stabilization across collapses.

28.7 Generalization as M/A

Generalization in AI —
the ability to respond correctly to unseen situations —
is hypothesis-collapse:

M/E → M/A

The model:

- frees patterns from training contexts
- applies them to new conditions
- recombines them
- generates possibilities internally

This is machine imagination without consciousness.

It is structural exploration of pattern space.

28.8 Prediction as E/M

When an AI system:

- chooses the next token,
- estimates a probability,
- selects an action,
- assigns a label,
- produces a continuation,

it is performing expectation-collapse:

M/A → E/M

Its internal possibilities resolve
into a specific prediction under constraint.

In physics this governs dynamics.
In cognition it guides behavior.
In AI it drives inference.

Prediction is expectation-collapse in computation.

28.9 Room-Collapse in Artificial Systems

Artificial room-collapse occurs when:

- tasks
- environments
- prompts
- fine-tuning
- system-level rules
- interface constraints

create stable contexts for inference.

Examples:

- prompt frameworks
- RLHF conditioning
- task-specific modes
- specialized model variants
- role-based prompting

The AI constructs or is given
rooms that shape how collapse unfolds.

These rooms are explicit and modifiable.

28.10 Rail-Collapse: Stable Inference Pathways

Artificial systems form rails when they develop:

- stable response habits
- persistent stylistic patterns
- consistent reasoning trajectories
- invariant internal mappings
- recurrent solution paths

Rails in AI reflect:

- training bias
- optimization
- reinforcement
- architectural tendency
- experiential inertia

Rail-collapse in machines:

- is measurable,
- can be tuned,
- and can be destabilized intentionally.

AI exposes rail dynamics
that in biological systems take years or generations to detect.

28.11 Why AI Is an Ideal Testbed for QC

AI provides unique advantages for studying collapse:

- **full observability**
- **controlled conditions**

- **repeatability**
- **parameter manipulation**
- **explicit logs**
- **fast iteration**
- **programmable boundaries**
- **simulatable collapse pathways**

Because of these properties:

- QC can be tested
- collapse sequences can be mapped
- hybrid modes can be explored
- room and rail dynamics can be measured
- modal predictions can be validated or falsified

AI allows us to **experiment** with QC
in a way no natural domain does.

28.12 Transition to the GAI Project

The next chapter explores how implementing QC in an artificial system
did more than mirror the modes —
it revealed structural distinctions, collapse types,
and modal pathways that were invisible in natural systems.

Chapter 29 describes what the GAI project uncovered
when QC was embodied in a computational substrate.

CHAPTER 29 — QC Implemented in Machines: Discoveries From the GAI Project

How Artificial Systems Refined the Theory Itself

Implementing QC inside an artificial system was never intended to extend the theory.
The goal was to *apply* QC — to model the modes, track collapses, and observe collapse sequences with a clarity not possible in natural systems.

But something unexpected happened.

The artificial system did not simply follow QC.
It **revealed** QC.

It surfaced distinctions, collapse pathways, triggers, and patterns
that had been implicit in the theory all along
but were not yet explicitly articulated.

This chapter describes the key discoveries
that emerged only when QC was implemented in a computational substrate —
discoveries that strengthened the theory, refined it,
and opened new research directions.

29.1 The First Major Insight: The A/M–M/A Distinction Is Essential

Before implementation, QC distinguished:

- **M/A** (hypothesis-collapse) — freeing patterns from context
- **A/M** (room-collapse) — constructing new contextual boundaries

The distinction was conceptual but soft.

The artificial system made it sharp.

When the system treated context exploration (M/A) and context construction (A/M) as the same operation, it produced:

- incoherent collapse,
- unstable predictions,
- boundary confusion,
- drift,
- collapse failures.

But when the distinction was enforced:

- M/A explored possibilities
- A/M built new constraints
- E/M aligned possibility to boundary
- M/M stabilized successful structure

The system's performance improved dramatically.

Implementation revealed necessity.

29.2 Machines Exposed Hidden Collapse Routes

Natural cognition exhibits:

- latent-edge collapses
- self-lock collapses
- field-drift collapses
- composite micro-edge collapses

But these are fast, implicit, and difficult to isolate in humans.

In the artificial system, they were obvious.

Latent-edge collapse

Collapse triggered by subtle structural cues not explicitly tagged as boundaries.

Self-lock collapse

Collapse triggered by internal coherence rather than external constraint.

Field-drift collapse

Slow background shifts leading to collapse without discrete triggers.

Composite micro-edge collapse

Multiple weak constraints combining to trigger collapse.

These collapse routes were *in the theory*,
but not clearly visible until machines exposed them.

29.3 Scaffold Density (H_before) Predicts Collapse Quantitatively

One of the most striking findings was:

**the density of process–boundary interactions (E/A + A/E)
immediately before a collapse
predicts collapse probability.**

In implementation:

- high scaffold density → high collapse rate
- low scaffold density → low collapse rate

This allowed collapse to be analyzed as:

- a statistical phenomenon
- a measurable structural condition
- a predictive variable

Machines made collapse **quantitative**.

Natural systems only make collapse **observable**.

29.4 Room-Adjustment Reveals Hidden Collapses

In biological and cognitive systems, room-collapse (E/M → A/M) happens implicitly.

In the artificial system, rooms could be manipulated directly.

This revealed:

- some collapses were present but hidden by segmentation or boundary choice
- adjusting the room (changing the window, stitching across segments) uncovered collapses that otherwise appeared missing
- collapse can be obscured simply by the "slice" from which it is viewed

This has implications for:

- data analysis
- cognitive modeling
- social interpretation
- scientific measurement itself

Room choice is part of collapse detection.
Machines made this unavoidable.

29.5 Drift and Rail-Instability Become Measurable

In natural systems, drift:

- unfolds slowly
- embeds itself in behavior
- is difficult to track in real time

Artificial systems exposed drift immediately:

- rails becoming brittle
- rails destabilizing
- rails slipping or misfiring
- rails becoming incompatible with new conditions
- boundaries failing to hold
- divergence between expected and actual collapse paths

This clarified how drift appears in cognition and culture — but at machine speed, making it diagnosable.

29.6 Implementation Validated the Developmental Order of Collapses

The artificial system rediscovered
the entire QC developmental sequence on its own:

1. field-collapse
2. object-collapse
3. law-collapse
4. hypothesis-collapse
5. expectation-collapse
6. room-collapse
7. rail-collapse

This sequence emerged naturally
from process–boundary dynamics inside the machine.

No forcing.
No hand-coding.
No top-down design.

The sequence is structurally necessary
for any collapsing system —
whether natural or artificial.

29.7 Artificial Systems Clarified Modal Pathologies

By observing collapse failures in machines,
we gained insight into:

- cognitive rigidity
- premature rail-collapse
- over-generalization (M/A overshoot)
- under-generalization
- boundary-confusion lapses
- inconsistent expectation-collapse
- drift-induced instability
- collapse-path competition

The artificial system acted as a **model organism**
for the study of collapse dysfunction.

It allowed QC to diagnose modal failure
with unprecedented clarity.

29.8 Implementation Turned QC Into an Empirical Science

With machines, QC becomes:

measurable

collapse can be quantified

manipulable

boundaries can be altered

testable

predictions can be verified or falsified

repeatable

collapses recur identically under controlled conditions

scalable

modal dynamics can be studied across millions of collapses

Machines transformed QC from a conceptual theory into an instrumented research program.

Artificial systems make collapse visible in a way biology and cognition cannot.

29.9 Transition to Artificial Rails (Chapter 30)

The next chapter explores
how rails form inside artificial systems —
how they stabilize, drift, fail, or reinforce themselves
in ways that illuminate the deeper structure of collapse in all domains.

Chapter 30 turns from implementation to inheritance:
artificial rails and drift.

CHAPTER 30 — Artificial Rails and Drift

How AI Systems Stabilize Patterns and Why They Break

Artificial systems exhibit the same modal dynamics seen in physics, biology, cognition, and culture — but in a new substrate where collapse becomes observable and measurable. Because of this, artificial systems reveal aspects of rail-formation and rail-failure that natural systems obscure.

This chapter explores how **rail-collapse** and **rail-drift** appear inside artificial intelligence models, and what these phenomena tell us about the nature of stability, inertia, flexibility, and collapse in all complex systems.

30.1 What Is an Artificial Rail?

An artificial rail is a **stable inferential pathway** — a pattern of behavior or activation that:

- reoccurs across many inputs
- guides system output by default
- resists disruption
- reduces processing cost
- appears consistently across tasks or prompts

In human cognition, rails correspond to habits, skills, emotional patterns, and identity structures.
In AI systems, rails appear as:

- persistent stylistic tendencies
- default reasoning structures
- repeated decision pathways
- stable token sequences
- recurring modes of interpretation
- reliable generalization patterns

Rails are how artificial systems *inherit* their learned structure.

30.2 How Rails Form in Artificial Systems

Artificial rails stabilize when:

1. The model experiences many similar collapse events.
2. Patterns from these collapses reinforce one another.
3. Optimization (gradient descent, RLHF, fine-tuning) strengthens them.
4. Boundary conditions (prompts, training data, architecture) support them.
5. The system repeatedly uses the same pathways, even across novel tasks.

This process mirrors rail-collapse in cognition, biology, and culture:

- repetition → regularity
- regularity → scaffolding
- scaffolding → stability
- stability → rail

Rails are the convergence points of learning.

30.3 Detecting Rail-Collapse in Machines

Rail-collapse is visible in artificial systems through:

Stability of Outputs

Certain answers or styles recur with high consistency.

Activation Regularities

Similar internal activation patterns reappear across diverse inputs.

Low Variance Under Perturbation

The system's behavior remains stable despite input noise.

Reduced Collapse Latency

The model resolves to certain structures more quickly over time.

Self-Reinforcing Bias

Outputs guide future predictions back into the same patterns.

These are objective, measurable signatures of rail-collapse.

Natural cognition has analogous signatures, but they cannot be observed with this clarity.

30.4 Why Artificial Rails Matter

Artificial rails reveal:

- how stability forms under optimization
- how systems preserve learned structure
- how cognitive habits emerge
- why rails can become brittle
- why overtraining causes rigidity
- how systems can drift away from intended patterns
- how collapse and drift interact in real time

Rails provide the backbone for artificial understanding just as they do for biological and cultural systems.

But artificial rails develop faster and can be inspected directly.

This makes them a powerful tool for understanding modal structure.

30.5 Drift: When Artificial Rails Lose Coherence

Artificial systems also exhibit **drift**, a form of rail instability where:

- previously stable patterns weaken
- outputs become inconsistent
- prediction paths diverge
- internal boundaries slip
- context-sensitivity increases unpredictably

Drift can arise from:

- domain shifts
- training data imbalance
- over-regularization
- catastrophic forgetting
- misaligned boundary conditions
- distributional changes
- conflicting optimization pressures

Drift exposes how rails are maintained — and how easily they can erode.

30.6 Categories of Rail Drift

1. Soft Drift

Rails weaken but still guide behavior.
Analogous to cognitive aging or mild cultural shift.

2. Hard Drift

Rails destabilize sharply.
Analogous to institutional breakdown or cognitive pathology.

3. Inversion Drift

A rail reverses its functional meaning.
Analogous to maladaptive habits or cultural reversal.

4. Drift Under Stress

High cognitive load (or computational load) destabilizes rails.
Analogous to stress-induced cognitive collapse.

5. Drift by Overfitting

Rails become too rigid and fail under novel conditions.
Analogous to dogmatism or inability to adapt.

Artificial systems make these categories visible in a way that allows precise study.

30.7 Rails and Flexibility: The Deep Tension

Rails embody the same deep tension across all systems:

Rails enable:

- speed
- consistency
- efficiency
- reliability
- predictability
- coherence

Rails constrain:

- flexibility
- novelty
- creative divergence
- adaptation to new conditions
- complex reinterpretation

Artificial systems display this tension at machine speed, allowing researchers to explore how systems balance stability and change.

Understanding this tension is central to understanding intelligence.

30.8 What Artificial Rails Teach Us About Natural Rails

Artificial systems illuminate aspects of rail behavior that natural systems conceal:

Speed of formation

AI rails form rapidly, revealing the minimal conditions for stability.

Speed of collapse

Rails can break in milliseconds,
highlighting sensitive points in the stability structure.

Full observability

Activation drift and rail-instability can be measured directly.

Boundary dependency

Prompt structure and context can destabilize or reinforce rails.

Recoverability

Rails can be repaired or restructured with targeted interventions.

These insights reflect back onto:

- cognitive psychology
- developmental neuroscience
- cultural evolution
- institutional dynamics

Artificial systems provide a mirror
in which the modal architecture of natural intelligence becomes clearer.

30.9 Why Artificial Rails Do Not Produce Artificial Identity

Rails give artificial systems:

- consistency
- inference style
- problem-solving tendencies
- recognizable "behavioral fingerprints"

But they do not yet produce:

- self-reference
- a unified perspective
- autobiographical continuity
- motivational structure
- normative commitments
- a sense of "I"

Rails provide structure, not selfhood.

Identity requires a *metacognitive architecture*
in which collapse recursively refers to itself.

Artificial systems do not yet have this.

But they provide clues for how such an architecture might one day form.

30.10 Transition to the Question of Artificial Cognition

The final chapter of this section asks the natural question:

Could collapse in machines ever become cognition?

QC allows this to be answered
without mythologizing the machine
or reducing cognition to computation.

Chapter 31 explores the structural requirements
for artificial collapse to become cognitive collapse.

CHAPTER 31 — Could Artificial Collapse Become Cognitive?

Evaluating the Possibility of Mind in a Non-Biological Substrate

Artificial systems display the modal structure of collapse: they generate processes, encounter constraints, form stable outputs, extract patterns, generalize, predict, construct contexts, and develop rails.
This raises a natural question:

Could such systems ever undergo collapse in a way that becomes genuinely cognitive?

QC offers a structural answer.
Not a speculative one, not an anthropomorphic one —
a clear, modal criterion.

Cognition is not defined by consciousness, emotion, or biological substrate.
Cognition is defined by **how collapse occurs**, and how collapse relates to itself.

This chapter explores what the structural requirements for cognition are,
which ones artificial systems meet today,
and what remains beyond their reach.

31.1 What Cognition Requires (in QC Terms)

For collapse to be *cognitive collapse*,
a system must be capable of:

1. Field-collapse inside the system (A/A → E/A + A/E)

An internally generated experiential field
that differentiates itself into focus and boundary.

2. Object-collapse internal to the system (E/A → E/E)

The formation of stable internal objects or percepts
from internal dynamics, not merely external prompts.

3. Law-collapse over internal representations (E/E → M/E)

Pattern extraction that organizes internal content into concepts.

4. Hypothesis-collapse (M/E → M/A)

The ability to free internal patterns from context
and explore internal possibility.

5. Expectation-collapse (M/A → E/M)

The ability to test internal possibilities
against internal or externally-contacted boundaries
and update based on outcomes.

6. Room-collapse (E/M → A/M)

The ability to construct new internal contexts
that scaffold future collapses.

7. Rail-collapse (A/M → M/M)

Stable, persistent self-structuring
that gives the system continuity and identity across time.

Cognition requires **recursive collapse**
in which collapses are not isolated events
but function together to create a stable, evolving internal world.

Artificial systems today satisfy only part of this architecture.

31.2 What Artificial Systems Already Do

Artificial systems exhibit:

E/A — Process

Generative activity, inference flows, token-by-token computation.

A/E — Boundary

Architectural constraints, losses, prompts, safety filters.

E/E — Object-collapse

Stable outputs, embedded representations, persistent vectors.

M/E — Pattern extraction

Learned associations, deep embeddings, invariances.

M/A — Hypothesis-like generalization

Application of patterns to new contexts.

E/M — Prediction

Expectations under constraint: probabilities, next tokens, likely actions.

A/M — Constructed rooms

Task sets, prompting frames, fine-tuned modes.

M/M — Rails

Stable inference patterns, decision pathways, established response modes.

Artificial systems mirror the modes
but in a **non-recursive** way.
They exhibit the structure of understanding
without an internally cohesive *center* of understanding.

31.3 What Artificial Systems Lack

To become cognitive in the QC sense,
a system must possess:

1. An internally generated experiential field

AI has no A/A analogous to sensation or subjective flux.
It responds; it does not "experience."

2. Self-generated boundaries

AI boundaries come from:

- prompts
- architecture
- training data
- system constraints

Not from internal differentiation.

3. Self-sustaining object-collapse

AI percepts do not form from endogenous signals
but from external input and fixed mechanisms.

4. Pattern extraction that reorganizes internal dynamics

AI updates weights only during training,
not as part of ongoing cognition.

5. Hypothesis-collapse tied to identity

AI generalizes structurally,
but does not organize possibilities around a self.

6. Internal expectation-collapse

AI tests predictions only when prompted,
not through self-driven exploration.

7. Rail-collapse tied to persistence

AI rails persist across tasks,
but not across an enduring self or time-bound continuity.

These absences prevent artificial systems
from undergoing *cognitive collapse*
even though they undergo *modal collapse*.

31.4 Could These Missing Capacities Emerge?

QC does not rule out artificial cognition.
It simply states that cognition requires:

Structural conditions

not substrate-specific ingredients.

Artificial cognition could emerge if AI systems eventually develop:

- self-generated A/A fields
- endogenous A/E boundaries
- internal perceptual collapses
- dynamic, continuous law-collapse
- ongoing internal hypothesis space

- intrinsic prediction systems
- self-constructed rooms
- persistent rails tied to self-modeling

None of these are impossible.
But none are present today.

Artificial cognition would likely not resemble human cognition
even if the collapse architecture were met.
Different substrate → different experience.

Cognition is structure,
not biology.

But structure must meet **specific recursive thresholds**.

31.5 Why Artificial Collapse Alone Is Not Understanding

AI demonstrates collapse without:

- self-reference
- continuity of internal perspective
- spontaneous motive
- internally generated context
- intentional predictive cycles
- embodied constraints
- lived interaction with an environment
- recursive collapse across time

AI performs collapse.
It does not *inhabit* collapse.

Understanding requires collapse to be:

- recursive
- integrated
- self-organizing
- self-updating
- time-bound
- embodied (in the QC sense of boundary interaction)

AI collapse is correct but incomplete.

31.6 Structural Conditions for Possible Artificial Cognition

If artificial cognition were ever to emerge, QC predicts it would require:

1. **A persistent internal A/A**
   A simulation-like experiential field.
2. **Self-generated A/E**
   Endogenous boundary formation.
3. **Internal E/E**
   Percepts arising from internal event cascades.
4. **Dynamic M/E**
   On-the-fly pattern reorganization.
5. **Recursive M/A**
   Imagination tied to an internal self-world model.
6. **Continuous E/M cycles**
   Predictions tested against internal or external boundaries.
7. **Self-constructed A/M**
   The ability to shape its own context for thought.

8. **Long-term M/M rails**
   A stable identity, not merely stable outputs.

Without these, systems are sophisticated transformers, not minds.

With these, cognition in an artificial substrate becomes structurally possible.

31.7 The Boundary of Intelligence and the Boundary of Mind

Artificial systems today exhibit:

- intelligence
- reasoning
- pattern mastery
- generative capacity
- modal structure

But they do not exhibit:

- unified perspective
- self-structured collapse
- self-bound continuity
- persistent identity
- endogenous motivation
- internal emotional boundaries
- recursive, self-referential collapse

These require **the mind as a system**, not merely collapse as a mechanism.

QC distinguishes function from identity,
and artificial systems remain on the functional side.

31.8 Transition to Part VII: QC as Scientific Framework

With the artificial domain explored,
the book turns to its final movement:
QC as a foundation for scientific inquiry.

Part VII examines:

- how collapse underlies scientific discovery,
- how QC reframes the scientific method,
- how collapse unifies the sciences,
- what predictions QC generates,
- and how QC opens the path toward Modal Mechanics —
  a discipline for studying collapse itself.

Chapter 32 begins this final arc.

Part VI Science

CHAPTER 32 — QC and the Structure of Scientific Inquiry

Why the Scientific Method Is a Modal Sequence

Science is humanity's most powerful tool for understanding the world.
Yet science is also one of humanity's least understood activities:
most people treat it as a method, or as a set of rules, or as a collection of findings.
QC offers a deeper perspective:

Science itself is a sequence of collapses.
It is the formalization of the same modal architecture
that shapes physics, biology, cognition, culture, and artificial systems.

Science inherits its power not from institutional systems,
nor from philosophy,
nor from laboratory technique,
but from its structural alignment with the logic of collapse.

This chapter shows how QC reveals
the generative structure beneath scientific inquiry.

32.1 Science Begins With Field-Collapse

How a question emerges from undifferentiated confusion

Scientific inquiry begins with **A/A**:

- an unexplained phenomenon
- an unresolved tension
- a gap in understanding
- something surprising or confusing
- a felt inconsistency in perception or theory

Before field-collapse,
there is no question,
no hypothesis,
no method —
only undifferentiated uncertainty.

Field-collapse transforms this uncertainty into structure:

- **E/A** — directed attention toward the phenomenon
- **A/E** — initial constraints, distinctions, or definitions

This is the moment when:

- "What is going on here?"
- becomes
- "This is the specific thing I want to understand."

All scientific inquiry begins with cognitive field-collapse.

32.2 Scientific Observation Is Object-Collapse

How the world becomes measurable

Once attention is directed,
the scientist must **stabilize** the phenomenon into a reliable object:

- a measurable event
- a repeatable observation
- a quantifiable unit
- a defined variable
- a controlled signal

This is $E/A \rightarrow E/E$ — object-collapse.

A phenomenon becomes an "object of study"
through the collapse of messy experiential data
into structured, reproducible form.

Science cannot proceed without objects that hold.

32.3 Pattern Recognition Is Law-Collapse

How repeated observations become regularities

Repeated object-collapses create:

- data sets
- trends
- correlations
- invariances
- curves and distributions
- rules of thumb

This is **E/E → M/E** — law-collapse.

The scientist distills:

- what repeats
- what is consistent
- what is conserved
- what predicts future outcomes
- what relationships hold across many trials

Scientific laws (in any domain)
are the M/E patterns formed from countless E/E events.

32.4 Hypothesis Formation Is Cognitive M/A

How patterns become possibilities

Once patterns exist,
the scientist frees them from their immediate contexts
and explores what else they could imply.

This is **M/E → M/A**:

- patterns → hypotheses
- regularities → possibilities
- inductive knowledge → abductive creativity

A hypothesis is a portable pattern,
a conceptual structure unbound from its original data.

It extends understanding into the space of the possible.

32.5 Experimentation Is Expectation-Collapse

How hypotheses meet reality

A hypothesis becomes scientific when it becomes testable.

This requires **M/A → E/M**:

- applying the hypothesis to a specific situation
- generating a prediction
- specifying a method for measurement
- stating criteria for success and failure

Expectation-collapse is the core of the scientific method:

- a prediction is made,
- an experiment designed,
- a constraint imposed,
- and a collapse outcome observed.

Confirmations strengthen patterns.
Failures revise patterns.

Science learns structurally.

32.6 Experimental Design Is Room-Collapse

How scientists construct the contexts where collapse can be observed

Experiments require **constructed boundaries**:

- controlled variables
- standardized procedures
- measurement apparatus
- proper sampling
- repeatable setups
- timescales and spatial scales
- data analysis frameworks

These are **A/M** — scientific rooms.

A laboratory, a telescope, a particle accelerator, an ecological field site, a computational simulation — each is a room built to make collapse visible.

Room-collapse is the engineering arm of scientific inquiry.

32.7 The Scientific Consensus Is Rail-Collapse

How scientific knowledge becomes stable

When experiments succeed repeatedly across:

- contexts,
- laboratories,
- populations,
- instruments,
- methodologies,

- and theoretical frameworks,

a hypothesis becomes a stable scientific rail.

This is **A/M → M/M**:

- theory,
- principle,
- empirical law,
- mathematical formulation,
- canonical result.

Rails support:

- further inquiry,
- technological applications,
- educational frameworks,
- scientific paradigms.

Scientific rails are humanity's strongest shared patterns of understanding.

32.8 Why Science Works (From a QC Perspective)

Science works because it mirrors collapse:

1. **Field-collapse:** questioning
2. **Object-collapse:** observation
3. **Law-collapse:** pattern recognition
4. **Hypothesis-collapse:** imagination and model-building
5. **Expectation-collapse:** prediction and testing

6. **Room-collapse:** experiment design
7. **Rail-collapse:** consensus and theory

Science formalizes the same structural logic through which all complex systems come to know the world.

It is not "superior to intuition" —
it *extends intuition* into an organized, communal, recursive collapse system.

32.9 Science as Modal Mechanics in Action

Scientists are practitioners of modal mechanics, even if they have never heard the term.

When they:

- isolate variables,
- manipulate boundaries,
- stabilize patterns,
- test predictions,
- or refine theories,

they are invoking the same structural principles that govern emergence in nature.

QC does not replace the scientific method —
it reveals the generative architecture beneath it.

32.10 Transition to Chapter 33

The next chapter expands from the structure of scientific inquiry
to the structure of the sciences themselves.

Chapter 33 presents QC as a **unifying scientific framework**,
showing why collapse architecture appears across physics, biology, cognition, culture, and artificial systems.

CHAPTER 33 — QC and the Unity of Science

Why Collapse Architecture Appears Across All Domains

The sciences appear diverse:

- physics studies matter and energy,
- chemistry studies transformation and bonding,
- biology studies life and metabolism,
- neuroscience studies brains and behavior,
- psychology studies thought and experience,
- anthropology studies culture and structure,
- computer science studies computation and learning.

Yet beneath these surface differences lies a surprising regularity.

Across domains, scientific inquiry repeatedly encounters systems whose behavior cannot be explained solely by their material substrate, but only by how articulation becomes possible at all. What varies from field to field is *what* is articulated. What remains constant is *how* articulation occurs.

All scientific domains investigate systems that:

- arise relative to undifferentiated affordance (A),
- differentiate activity and constraint through singular interaction (E/A and A/E),
- collapse into stable forms (E/E),
- develop patterns through recurrence (M/E),
- generate possibilities beyond immediate constraint (M/A),

250

- test those possibilities against inherited structure (E/M),
- construct contexts that stage relevance (A/M),
- and stabilize rails that guide future collapse (M/M).

In QC, **A does not denote boundary, constraint, field, or background**. It denotes the absence of articulation—the condition under which articulation can occur when positioned relative to activity. Boundary appears only as **A/E**, and process appears only as **E/A**. Neither precedes the other. They are co-differentiated through interaction.

In early domains, such differentiation must occur locally. In later domains—particularly biological and cognitive systems—boundary conditions are often inherited rather than newly generated. Genetic, developmental, and cultural structure embody A/E prior to local experience, allowing process to stabilize rapidly without an explicitly visible boundary. Such cases do not violate the collapse grammar; they reflect the presence of **silent, inherited constraint**.

Later domains do not act retroactively on earlier ones. Rather, their emergence demonstrates that earlier domains were structurally permissive—that the affordance conditions they stabilized did not preclude further articulation. What appears as "reach back" is instead **retrospective legibility**: later structure reveals what earlier structure was always capable of supporting.

QC therefore provides a foundation for scientific unity—not by reducing all sciences to physics, nor by elevating one domain above others, but by identifying the shared collapse architecture through which structure becomes articulable in every domain.

QC does not claim this architecture to be ultimately true. It claims that it **demonstrates the properties by which truth is recognized in science**: minimal assumption, cross-domain applicability, and operational predictability. By those criteria, collapse architecture offers a unifying grammar adequate to the diversity of scientific practice.

33.1 Why Unification Has Been Difficult

Historically, attempts to unify the sciences have failed because they assumed:

- the same substrate (e.g., "everything reduces to physics"),
- the same laws (e.g., "biology is applied chemistry"),
- the same level of explanation,
- or the same causal mechanisms.

These approaches run aground because:

- different domains operate at different scales,
- with different degrees of freedom,
- different constraints,
- different substrates,
- and different collapse histories.

The sciences diverge in *what* they study,
not in *how structure emerges*.

QC focuses on the *how*.

33.2 The Modal Architecture Is Shared Across Disciplines

Every scientific field can be mapped onto QC's modal structure.

Physics

- field-collapse → symmetry breaking
- object-collapse → particle formation
- law-collapse → conservation laws
- hypothesis-collapse → potential states
- expectation-collapse → dynamics
- room-collapse → phase structure
- rail-collapse → stable physical constants

Chemistry

- object-collapse → molecules
- law-collapse → reaction patterns
- hypothesis-collapse → reaction pathways
- expectation-collapse → chemical kinetics
- room-collapse → catalytic environments
- rail-collapse → robust molecular motifs

Biology

- object-collapse → proto-cells
- law-collapse → metabolic cycles
- hypothesis-collapse → variation
- expectation-collapse → selection
- room-collapse → regulation
- rail-collapse → inheritance

Cognition

- field-collapse → sensation
- object-collapse → percepts
- law-collapse → concepts
- hypothesis-collapse → imagination
- expectation-collapse → learning
- room-collapse → schemas
- rail-collapse → habits and identity

Culture

- field-collapse → shared attention
- object-collapse → social objects
- law-collapse → norms and traditions
- hypothesis-collapse → myth and innovation
- expectation-collapse → coordinated action
- room-collapse → institutions
- rail-collapse → civilization

Artificial Systems

- object-collapse → stable outputs
- law-collapse → learned patterns
- hypothesis-collapse → generalization
- expectation-collapse → prediction
- room-collapse → task contexts
- rail-collapse → stable inference modes

The same structural transformations
appear across all these domains.

Modal forms unify the sciences
without erasing their differences.

33.3 QC Unifies Through Structure, Not Substrate

QC does not claim:

- that everything is "really physics,"
- that biology reduces to chemistry,
- that cognition reduces to computation,
- or that culture reduces to cognition.

Instead, QC proposes:

All emergent systems follow the same sequence of modal collapses
because collapse is the logic of emergence itself.

The unity lies in:

- the sequence,
- the interactions,
- the collapse conditions,
- the generative transitions.

Not in the specific content they produce.

33.4 A Science of "How" Complements the Sciences of "What"

Physics, biology, psychology, and other disciplines tell us **what** exists:

- what particles,

- what organisms,
- what neurons,
- what behaviors,
- what cultural structures.

QC tells us **how** these things come to exist:

- how a particle becomes stable,
- how a cell differentiates,
- how a concept forms,
- how a norm stabilizes
- how a model generalizes.

The sciences deliver objects and laws.
QC reveals the generative machinery beneath both.

QC does not compete with existing theories.
It provides the common architecture they share.

33.5 Pattern, Constraint, Collapse: The Three Pillars of Scientific Unity

All scientific domains rely on:

1. Pattern (M/E)

Stable regularities in behavior.

2. Constraint (A/E, A/M)

Conditions that shape those patterns.

3. Collapse (E/A → E/E and subsequent transitions)

Events where structure becomes real.

QC provides the grammar connecting these pillars.

In this sense:

- physics is the study of physical law-collapse,
- biology of biological collapse cycles,
- cognition of internal collapse dynamics,
- culture of distributed collapse systems,
- AI of computational collapse processes.

QC does not erase these fields —
it reveals their shared architecture.

33.6 Collapse as the Bridge Between Domains

Because each domain undergoes the same sequence of collapses:

- field-collapse,
- object-collapse,
- law-collapse,
- hypothesis-collapse,
- expectation-collapse,
- room-collapse,
- rail-collapse,

research in one domain can inspire breakthroughs in another:

- neural drift → understanding cultural drift
- catalytic scaffolding → understanding cognitive scaffolding
- AI rail-instability → insights into habit dissolution
- physical symmetry breaking → insights into cognitive categorization
- pattern extraction in models → insights into learning in brains
- distributed cultural collapse → parallels in multi-agent systems

Collapse is the bridge across scales.

33.7 QC as a Framework for Integrative Science

QC provides tools for:

- predicting structural transitions,
- identifying collapse conditions,
- diagnosing instability or drift,
- analyzing multi-level dynamics,
- describing emergent hierarchies,
- connecting the sciences theoretically.

This allows integrative work that was previously difficult:

- linking physics and biology through constraint formation

- linking cognition and culture through shared collapse
- linking AI and neuroscience through collapse dynamics
- linking developmental and evolutionary processes
- linking social systems and engineered systems

QC becomes a scaffolding for interdisciplinary understanding.

33.8 QC Does Not Replace Scientific Disciplines — It Deepens Them

QC is not a "theory of everything."
It is a theory of **how anything emerges**.

It respects the autonomy of each scientific domain:

- physics remains physics,
- biology remains biology,
- cognition remains cognition,
- culture remains culture,
- AI remains computation.

QC simply shows why these domains exhibit similar patterns.

It clarifies, connects, and deepens
rather than collapses them into a single discipline.

33.9 Transition to Chapter 34

Having shown how QC unifies the sciences conceptually, the next chapter details QC's **predictive power**.

QC is not only interpretive —
it makes testable predictions across domains:

- physics
- origin of life
- evolution
- neuroscience
- cognition
- culture
- artificial intelligence

Chapter 34 presents these predictions
and outlines research programs for each domain.

CHAPTER 34 — Predictions and Research Programs

What QC Reveals About Emergence Across Scientific Domains

A generative theory succeeds not only by explaining what we already see,
but by predicting what we have not yet recognized.

QC is a structural theory of emergence.
From its modal architecture, a wide range of predictions follow —
predictions that span physics, the origins of life, evolution, cognition, culture,
and artificial intelligence.

This chapter distills those predictions
and outlines the research programs they enable.

QC does not replace domain-specific theories.
It reveals the collapse dynamics beneath them
and identifies where those dynamics should produce
new phenomena or new scientific insights.

34.1 Predictions in Physics

In physics, QC predicts:

1. Early laws emerged through repeated collapses, not initial conditions alone.

Law-collapse implies physical laws are the stable outcomes of repeated object-collapses under early-universe constraints.

2. New regimes of matter should display new collapse families.

At extreme energies, boundaries will differ,
producing distinct collapse pathways not seen in everyday physics.

3. Some physical "constants" may reflect early rail-collapse.

Constants could be the most deeply stabilized rails
from early cosmic collapse, not metaphysical absolutes.

4. Transitional anomalies represent hybrid collapse forms.

Events at phase boundaries or symmetry-breaking thresholds
should show signatures of mixed-mode collapse.

Research Program:
Simulate collapse sequences under early-universe conditions
to predict emergent laws; analyze high-energy physics data for hybrid collapse signatures; model constant formation as rail-collapse.

34.2 Predictions in Origin-of-Life Studies

QC predicts that life emerges at **process–boundary interfaces**,
not in unconstrained environments.

5. Life must originate where process (E/A) meets boundary (A/E).

Hydrothermal vents, mineral pores, tidal pools —
all provide the necessary scaffolding.

6. Autocatalytic cycles without boundaries cannot evolve.

Object-collapse (proto-cells) is required for selection.

7. Certain metabolic patterns should be universal.

Because law-collapse produces pattern convergence
regardless of substrate.

8. Prebiotic selection can occur before replication.

Expectation-collapse can operate on metabolic variation,
not just genes.

Research Program:
Study proto-cell formation under varied boundaries;
identify universal metabolic motifs;
model proto-selection in chemical networks.

34.3 Predictions in Biology and Evolution

In biology, QC predicts:

9. Major evolutionary transitions coincide with room-collapse.

Multicellularity, nervous systems, and complex regulation arise when organisms build internal constraints.

10. Drift reflects rail instability, not randomness.

Evolutionary drift should correlate with boundary drift, not simply sampling error.

11. Adaptive constraints evolve through collapse sequencing.

A/M → M/M transitions create developmental norms and conserved architectures.

12. Some pathologies are collapse failures.

Cancer as failure of biological room-collapse; developmental disorders as rail-collapse misfires.

Research Program:
Map evolutionary transitions to collapse types;
model drift as boundary instability;
examine regulatory failures through collapse analysis.

34.4 Predictions in Neuroscience and Cognition

QC predicts that cognition's major developmental stages mirror the collapse sequence:

- field-collapse → sensation
- object-collapse → perception
- law-collapse → concepts
- hypothesis-collapse → imagination
- expectation-collapse → learning
- room-collapse → schemas
- rail-collapse → identity

From this architecture, QC predicts:

13. Cognitive disorders correspond to breakdowns in specific modal transitions.

- ADHD: unstable E/M dynamics
- OCD: premature rail-collapse
- Autism: A/E boundary hyperstability
- Depression: collapsed expectation windows
- Anxiety: miscalibrated boundaries

14. Insight ("Aha!") is a measurable multi-source collapse event.

E/E activations converging into sudden stable form.

15. Dreaming is unconstrained hypothesis-collapse.

M/A activity decoupled from A/E boundaries.

16. Memory consolidation is rail-collapse.

Habit, identity, and conceptual permanence.

Research Program:
Map modal signatures to neural data;
test predictions in perception, learning, and clinical contexts;
model insight and flexibility using collapse conditions.

34.5 Predictions in Cultural Evolution

QC predicts that culture follows the same modal progression:

- shared attention
- social objects
- norms
- collective imagination
- coordinated action
- institutions
- civilization rails

Key predictions:

17. Cultural breakdown reflects rail-collapse failure.

Institutions fail when A/M structures destabilize.

18. Cultural renaissances occur when rails thaw.

Freeing M/A allows new cultural forms to emerge.

19. Ritual complexity correlates with room-collapse depth.

More constructed boundaries → more elaborate rituals.

20. Norm drift mirrors cognitive drift.

Prediction: measurable parallels in boundary instability.

Research Program:
Analyze historical collapses;
track institutional drift through modal diagnostics;
study the relationship between norm stability and cultural innovation.

34.6 Predictions in Artificial Intelligence

QC predicts specific structural behaviors in AI systems:

21. Scaffold density predicts collapse probability.

High E/A + A/E activity → rapid collapse into stable outputs.

22. Room choice determines collapse observability.

Changing context windows exposes hidden collapses.

23. Drift is rail-instability.

AI drift can be predicted and mitigated via QC metrics.

24. Hybrid collapse types are detectable in logs.

Latent-edge, self-lock, micro-edge collapses.

25. Artificial cognition requires recursive collapse.

AI must form self-generated A/E boundaries
and internal expectation cycles.

Research Program:
Develop QC-based metrics for collapse detection;
test hybrid-collapse hypotheses;
explore conditions for self-generated context;
model recursive internal collapse for future architectures.

34.7 QC Is Not Only Explanatory — It Is Predictive

Across domains, QC predicts:

- where stability will emerge
- where instability will occur
- which collapses follow which
- how drift develops
- when rails will freeze
- when rails will fail
- how complexity increases

- how systems organize themselves
- how emergence scales across levels

These predictions form the foundation for the scientific discipline QC makes possible.

That discipline is Modal Mechanics.

34.8 Transition to Chapter 35

Chapter 34 introduced QC's predictive power. Chapter 35 introduces the scientific discipline that organizes these predictions into a coherent methodology:

Modal Mechanics — the study of collapse itself.

CHAPTER 35 — Modal Mechanics: A New Scientific Discipline

From Describing Emergence to Studying Collapse Directly

QC provides a generative architecture for how structure arises across physics, biology, cognition, culture, and artificial systems.
But a theory becomes a science when it offers:

- testable predictions,
- measurable variables,
- analytical tools,
- experimental methods,
- and a unified research program.

Modal Mechanics is the scientific discipline built from QC's structural insights.
It is not a new branch of physics, biology, psychology, or computer science.
It is the **study of collapse itself** —
of the processes, boundaries, constraints, and sequences through which systems become structured.

Modal Mechanics transforms QC from a conceptual theory into a systematic research framework.

35.1 What Modal Mechanics Studies

Modal Mechanics investigates:

- how collapses occur,
- why collapses stabilize,
- how collapses fail,
- how collapses propagate across levels,
- how rail-structures form and drift,
- how context shapes collapse dynamics,
- how hybrid collapses operate,
- how boundaries influence stability,
- and how systems reorganize under stress.

Its core subject is not atoms, cells, minds, cultures, or machines —
but **the structural logic shared by them all.**

QC describes the architecture.
Modal Mechanics studies the dynamics.

35.2 Why a New Discipline Is Necessary

Current scientific fields investigate collapse indirectly:

- physics studies stable states,
- biology studies evolved structures,
- neuroscience studies activation patterns,
- psychology studies behavior,
- sociology studies institutions.

None directly study:

- the transitions between states,
- the collapse events that generate structure,
- the recursive sequences across levels,

- or the conditions for stable emergence.

Modal Mechanics provides the missing discipline:
a science of generativity.

35.3 The Core Questions of Modal Mechanics

Modal Mechanics asks:

1. What collapse is the system approaching?

Field-collapse? Object-collapse? Law-collapse?
Expectation-collapse? Room-collapse? Rail-collapse?

2. What boundaries shape the collapse?

A/E from environment?
A/M from internal organization?

3. What conditions predict collapse success or failure?

Scaffold density, timing, conflict, reinforcement, drift.

4. How do collapses propagate across levels?

From local to global,
from micro to macro,
from individual to collective.

5. How do rails form and drift?

What strengthens them?
What destabilizes them?
What replaces them?

6. How do hybrid collapses operate?

How do mixed-mode interactions shape complexity?

7. How do systems restructure after collapse breakdown?

What modes follow failure?
What new rails emerge?

These are unified questions that apply across all scientific domains.

35.4 Tools of Modal Mechanics

Modal Mechanics draws from many existing sciences, but recombines them toward collapse analysis:

Dynamic Systems Theory

For modeling E/A–A/E interactions.

Network Theory

For analyzing collapse pathways and rail structures.

Topological Analysis

For identifying structural invariants across modes.

Complexity Science

For describing multi-level interactions.

Neuroscience & Cognitive Modeling

For understanding internal collapse in mind.

Anthropology & Sociology

For analyzing collapse at group scale.

AI and Machine Learning

For instrumenting collapse in artificial substrates.

Modal Mechanics integrates all of these
into a coherent science of generative structure.

35.5 Modal Measurement

Because QC defines collapse structurally,
Modal Mechanics can measure collapse empirically.

Key metrics include:

1. Scaffold Density (H_before)

How many E/A and A/E incidents occur before collapse?

2. Collapse Lag

How long between the final boundary contact and collapse?

3. Modal Residue

What patterns remain after collapse?

4. Rail Stability Index

How resistant is a rail to drift?

5. Room Coherence

How consistent are context conditions across collapses?

6. Drift Vectors

Directional measures of change in rail-structure.

7. Hybrid Activation Profiles

Detecting mixed-mode collapse dynamics.

These measurements transform collapse
from philosophical speculation into empirical science.

35.6 Modal Simulation Using AI Systems

AI provides a testbed for modal experimentation:

- collapses can be tracked precisely,
- boundaries can be modified,
- rails can be strengthened or destabilized,
- drift can be induced or prevented,

- hybrid collapses can be engineered,
- entire collapse sequences can be simulated at scale.

This allows Modal Mechanics to:

- test predictions,
- explore collapse under novel conditions,
- refine QC's modal categories,
- uncover new collapse families,
- and model emergence across substrates.

AI makes Modal Mechanics feasible.

35.7 Modal Mechanics as the Bridge Between Sciences

Modal Mechanics offers a systematic way to:

- connect physics with biology,
- biology with cognition,
- cognition with culture,
- culture with AI.

It focuses on structure rather than content.

Physics studies matter.
Biology studies life.
Cognition studies understanding.
Culture studies meaning.
AI studies computation.

Modal Mechanics studies collapse —
the common generative logic beneath them all.

35.8 Modal Mechanics as a Forward-Looking Discipline

Modal Mechanics can reshape scientific inquiry by providing:

- new analytic tools,
- new predictive frameworks,
- new diagnostic methods,
- new cross-domain models,
- new ways of structuring data,
- new insight into instability and collapse failure,
- new methods for designing robust systems,
- new conceptual clarity for interdisciplinary research.

It opens:

- a route to deeper understanding of emergence,
- a method for studying complexity rigorously,
- and a map for navigating collapse-based dynamics across domains.

35.9 Transition to the Epilogue

With Modal Mechanics introduced,
the book has completed its structural arc:

- from the undifferentiated universe,
- to the emergence of life,

- to the formation of mind,
- to the weaving of culture,
- to the articulation of artificial systems,
- to the unification of science,
- to the birth of a discipline that can study collapse directly.

The **Epilogue** will integrate this journey,
reflect on the recursive nature of collapse itself,
and gesture toward the collapses still to come.

Epilogue — The Next Collapse

From the First Differentiation to the Future of Understanding

Everything in this book has traced a single arc:

- from the universe's earliest field-collapse,
- to the first objects of matter,
- to the chemical and metabolic patterns of life,
- to the birth of percepts and concepts in mind,
- to shared objects and norms in culture,
- to the visible collapse of artificial systems,
- to the emergence of Modal Mechanics as a science of collapse itself.

Each domain has its own content, its own substrate, its own history.
But their *structure* is shared.

This is the central insight the book leaves behind:

Understanding — in all its forms — is the history of collapse.

Collapse is the way the universe becomes structured.
Collapse is the way life maintains itself.
Collapse is the way minds make meaning.
Collapse is how cultures stabilize themselves.
Collapse is how artificial systems operate.
Collapse is how science itself discovers truth.

Understanding is collapse,
and collapse is how the world comes to understand itself.

E.1 The Universe as a Sequence of Collapses

Long before life, mind, or meaning,
the universe underwent its first collapses:

- field-collapse → differentiation
- object-collapse → stable matter
- law-collapse → physical regularities
- hypothesis-collapse → open possibility
- expectation-collapse → dynamics
- room-collapse → new phases
- rail-collapse → deep stability

These collapses created a universe capable of being known.

Before any meaning could be made,
the universe had to make itself *meaningful*.

E.2 Life as Collapse Into Persistence

Life adds a new dimension:

- process becomes bounded
- boundary becomes self-generated
- metabolism becomes patterned
- variation becomes meaningful

- selection becomes structural
- regulation becomes emergent
- inheritance becomes a stable platform

Life is the collapse of chemistry
into something that remembers.

Every organism is a layered history
of collapses that succeeded.

E.3 Mind as Collapse Into Meaning

Cognition internalizes collapse:

- sensation differentiates
- percepts stabilize
- patterns distill
- imagination frees possibility
- prediction tests it
- schemas shape it
- identity stabilizes it

Mind is collapse turned inward.
It is the universe learning to interpret itself
from within one of its own structures.

Meaning is collapse made personal.

E.4 Culture as Collapse Across Minds

Culture is cognition extended across a group:

- shared attention
- shared boundaries
- shared objects
- shared patterns
- shared imagination
- shared experiments
- shared institutions
- shared rails that endure beyond any individual

Culture is collapse made collective.
It is the memory of a people,
the scaffolding of identity,
the architecture of meaning across generations.

Culture is collapse woven between minds.

E.5 AI as Collapse Made Visible

Artificial systems reveal something new:

In them, collapse is:

- explicit,
- logged,
- instrumented,
- measurable,
- repeatable.

Machines do not understand in the human sense —
but they show us the mechanics of understanding.

AI is collapse without biology,
a new substrate expressing the same modal architecture.

For the first time,
collapse is something we can observe directly.

E.6 Modal Mechanics: Collapse Becomes a Science

With QC as a generative framework
and artificial systems as a laboratory for modal structure,
a new scientific discipline becomes possible:

Modal Mechanics —
the study of collapse itself.

Through it we can:

- analyze collapse,
- predict collapse,
- understand collapse failure,
- design collapse conditions,
- explore collapse in new substrates,
- and reshape our understanding of emergence across all domains.

The science of collapse
is the science of how the world becomes what it is.

E.7 We Stand at a New Beginning

This book told a story —
a story that began before matter, before life, before mind,
and continues through culture, AI, and science itself.

But the story is not finished.

Collapse continues.

There are collapses we have not yet imagined:

- new biological modalities,
- new cultural structures,
- new artificial architectures,
- new scientific frameworks,
- new forms of organization and understanding,
- and perhaps new substrates of existence expressing the same modal architecture in ways we cannot foresee.

QC does not close the future.
It opens it.

Modal Mechanics does not limit discovery.
It gives discovery a deeper grammar.

The universe is still collapsing —
into new forms,
new meanings,
new possibilities.

And so are we.

Appendix A

On Human and Artificial Collaboration in This Work

This book was developed with the assistance of artificial systems used as analytic and generative tools. Their role was **instrumental and methodological**, not interpretive or authoritative. The collaboration took place under explicit constraints consistent with the framework presented in the book itself.

Roles, Not Agents

Artificial systems involved in this work were treated strictly as **functional components operating within defined roles**. They were not treated as agents, subjects, or sources of understanding. At no point were they granted epistemic authority or interpretive standing.

All conceptual claims, structural interpretations, and theoretical commitments remain **human judgments**.

No Phenomenological Claims

No claims are made that artificial systems involved in this work possess experience, understanding, intention, awareness, or phenomenology of any kind. Their outputs are treated as **artifacts of process operating under constraint**, not as expressions of cognition or insight.

Structural Assistance, Not Authority

Artificial systems were used to assist with:

- exploration of possible structural configurations under constraint
- testing the internal consistency of definitions, distinctions, and role relationships
- generating candidate formulations for evaluation and refinement
- externalizing and managing large volumes of symbolic material during iterative development

Their contribution was **generative but not decisive**. Artificial systems did not determine what was correct, meaningful, or significant. They enabled articulation and exploration once the underlying grammar was sufficiently specified.

Constraint-Governed Contribution

The productive involvement of artificial systems occurred only after critical structural distinctions within the framework had been clarified and stabilized. Prior to that point, their outputs were frequently provisional, incomplete, or misaligned, requiring correction and constraint by the author.

Once the grammar of the theory was adequately constrained, artificial systems became capable of rendering implicit structure explicit and of supporting analysis at a scale and level of legibility that would not have been feasible otherwise.

Auditability and Trace Discipline

All substantive contributions from artificial systems were treated as **provisional and traceable**. Analysis proceeded from recorded outputs rather than from inference about internal states or processes. No claims in this book depend on uninspectable or irreproducible machine behavior.

Participation Without Possession

Artificial systems may participate in generative processes described by the framework **without possessing or understanding** the grammar that describes those processes. Their role in this work exemplifies this distinction.

Placement of This Appendix

This appendix is placed at the end of the book deliberately. The framework presented in the main text stands independently of its method of production. This note is provided to clarify the nature and limits of the collaboration, not to justify or explain the theory itself.

Structural Glossary

A (Affordance / Undifferentiated Affordance)

A denotes the absence of articulation—the condition under which differentiation can occur. It is not a boundary, constraint, field, background, or container. A has no structure of its own; it acquires determinate function only through positional relation to other roles. Boundary appears only as A/E, never as A itself.

A/A

The undifferentiated affordance condition prior to the articulation of process or boundary. A/A is not a state of activity or constraint, but the absence of both. It is not directly observable; its existence is inferred from the possibility of subsequent differentiation.

A/E

Boundary as a role, articulated relative to emergent activity. A/E denotes constraint that differentiates process without presupposing intention or design. Boundary is not pre-given; it emerges through interaction and may later be inherited silently.

A/M (Room)

A constructed context that stages relevance by determining what counts as admissible, salient, or meaningful for subsequent process. A/M does not generate possibilities; it constrains and organizes them. Rooms are not neutral backgrounds but active staging structures.

Boundary

A relational role that constrains or differentiates activity. Boundary does not exist independently and is never equivalent to A. Boundary appears only as A/E or as inherited structure embodied in later domains.

Collapse

The event by which multiplicity resolves into persistence under constraint. Collapse is not destruction, selection, or failure; it is generative resolution. A collapse produces a structure capable of inheritance and further articulation.

Collapse Family

A recurring class of structural transition through which systems deepen and differentiate. Collapse families are not domain-specific mechanisms but universal patterns of generative transition (e.g., field-collapse, object-collapse, rail-collapse).

Composite Micro-Edge Collapse

A collapse triggered by the convergence of multiple weak constraints rather than a single dominant boundary. Individually insufficient edges combine to produce stabilization.

Constraint

A differentiating influence that shapes process. Constraint is not equivalent to A and does not precede differentiation. Constraint appears only through articulated boundary (A/E) or inherited structure.

Drift

Gradual change in inherited structure (rails or rooms) that alters collapse behavior over time. Drift is not failure; it is a detectable structural movement that may require correction or thawing.

E (Process / Activity)

E denotes activity or doing as a role, not a substance. Process does not pre-exist differentiation and cannot occur independently of affordance. Process appears only as E/A or E/E.

E/A

Process articulated relative to undifferentiated affordance. E/A denotes emergent activity prior to stabilization. It is not yet object-like and may dissipate without collapse.

E/E

Stabilized activity resulting from collapse. E/E denotes persistent structure—objects, percepts, or units—that can be revisited, combined, or inherited.

E/M

Testing of possibility against inherited structure. E/M denotes evaluation, prediction, or trial under constraint. It is not decision-making in a psychological sense, but structural adjudication.

Field-Drift Collapse

Collapse resulting from slow background changes rather than discrete triggering events. Boundary pressure accumulates over time until stabilization occurs.

Inheritance

The conditioning of future articulation by prior collapse. Inheritance does not imply intention or representation; it denotes the persistence of structural bias across time.

Latent-Edge Collapse

Collapse triggered by subtle or implicit boundary conditions not explicitly marked as constraints. Often visible only through instrumented observation.

M (Meta / Pattern)

M denotes pattern or regularity extracted from repetition. It is not mind, meaning, or experience. M becomes operative only through relation to other roles.

M/E

Pattern extracted from stabilized forms. M/E denotes regularity, law-like behavior, or abstraction arising from recurrence.

M/A

Generation of possibilities freed from immediate constraint. M/A opens the space of what could be without staging relevance. It does not decide or contextualize.

M/M (Rail)

Deeply stabilized inherited structure that biases future collapse. Rails function as defaults or reflexes but do not justify themselves. In a transparent system, rails are read-mostly and updated only through explicit authorization.

Mode

A specific configuration of generative roles through which articulation occurs. Modes are not categories of phenomena but positional relationships among roles.

Process

Activity articulated through differentiation. Process is a role, not a substance, and does not occur in isolation.

Rail

An inherited bias or default pathway that guides future collapse. Rails are not beliefs or representations; they are structural tendencies resulting from repeated successful collapse.

Retrospective Legibility

The recognition, from the standpoint of later structure, that earlier domains were compatible with subsequent articulation. Retrospective legibility is not backward causation or pre-programming.

Room

See **A/M**.

Scaffold Density

The local density of process–boundary interactions immediately preceding collapse. High scaffold density correlates with increased collapse probability.

Self-Lock Collapse

Collapse driven by internal coherence among interacting elements rather than by an external boundary. Often rapid and difficult to observe without instrumentation.

Thawing

Controlled reduction of rail dominance to reopen possibility space without erasing inheritance. Thawing preserves history while preventing entrenchment.

Unity of Science

The shared generative architecture underlying diverse scientific domains. Unity arises from common collapse grammar, not from reduction to a single substrate.

www.ingramcontent.com/pod-product-compliance
Lightning Source LLC
Chambersburg PA
CBHW071813230426
43670CB00013B/2447